MICHIGAN ACADEMY OF SCIENCE, ARTS AND LETTERS

VOLUME II

CONTAINING PAPERS SUBMITTED AT THE ANNUAL MEETING IN 1922

THE MACMILLAN COMPANY
NEW YORK BOSTON CHICAGO
ATLANTA SAN FRANCISCO

MACMILLAN & CO., Limited
LONDON BOMBAY CALCUTTA
MELBOURNE

THE MACMILLAN CO. OF CANADA, Ltd.
TORONTO

PAPERS

OF THE

MICHIGAN ACADEMY OF SCIENCE ARTS AND LETTERS

EDITORS

PAUL S. WELCH

UNIVERSITY OF MICHIGAN

EUGENE S. McCARTNEY

UNIVERSITY OF MICHIGAN

New York

THE MACMILLAN COMPANY

LONDON: MACMILLAN & COMPANY, Limited

1923

Set up and printed,
December, 1923

Printed and bound by CPI Group (UK) Ltd, Croydon, CR0 4YY

Paperback ISBN: 978-0-472-75122-8

OFFICERS FOR 1922

President

A. F. SHULL, Ann Arbor

Vice-Presidents

AGRICULTURE, V. H. Church, East Lansing
ANTHROPOLOGY, C. H. Cooley, Ann Arbor
BOTANY, H. T. Darlington, East Lansing
ECONOMICS, W. A. Paton, Ann Arbor
GEOLOGY and GEOGRAPHY, W. I. Robinson, East Lansing
PSYCHOLOGY, C. H. Griffitts, Ann Arbor
SANITARY and MEDICAL SCIENCE, R. W. Pryer, Detroit
ZOÖLOGY, B. A. Barber, Hillsdale

Secretary

I. D. SCOTT, Ann Arbor

Librarian

PETER OKKELBERG, Ann Arbor

Editor

P. S. WELCH, Ann Arbor

CONTENTS

BOTANY

PAGE

The Genus Ribes with reference to the White Pine Blister Rust in the West. Dow Vawter Baxter 1

The Distribution and Pollination of Certain Sea-Grasses. H. H. M. Bowman 3

Notes on the Michigan Flora: Part V. Oliver Atkins Farwell 11

Extensions of Range of Plants in the Douglas Lake Region, Cheboygan County, Michigan. Frank C. Gates 47

Studies on the Hydrogen Ion Concentration of Plant Juices. F. G. Gustafson 49

The Genus Armillaria in the United States and its Relationships. C. H. Kauffman 53

Two Unreported Parasites of Hevea brasiliensis. Carl D. La Rue 69

A Leaf-Fall Disease of Hevea brasiliensis Muell.-Arg. due to Gloeosporium Alborubrum Petch. Carl D. La Rue and H. H. Bartlett 73

Diplodia Disease of Hevea brasiliensis. Carl D. La Rue and H. H. Bartlett 91

Stomatal Behavior of Plants in the Greenhouse in Winter. Hugh B. Smith 109

ECONOMICS

Foreign Investment and Foreign Trade. C. E. Griffin...... 119

The Revenue Act of 1921. F. E. Ross.................... 137

Certain Problems of Labor and the Public Attitude. Herbert F. Taggart.. 143

GEOLOGY

Lipsanocystis traversensis, A New Cystid from the Devonian of Michigan. G. M. Ehlers and J. B. Leighly...... 155

METEOROLOGY

Meteorological Summary, Douglas Lake, Michigan. Frank C. Gates... 161

Meteorological Data, Douglas Lake, Michigan. Frank C. Gates... 162

PAGE

PSYCHOLOGY

ILLUSIONS OF ORIENTATION. Nathan A. Harvey 165

SANITARY AND MEDICAL SCIENCE

SOME ASPECTS OF THE CHLORINATION TREATMENT AT THE FLINT FILTRATION PLANT. R. S. Buzzell 173
A SIMPLE QUANTITATIVE PRECIPITATION REACTION FOR SYPHILIS. R. L. Kahn 177
THE FILTRATION PLANT FOR DETROIT. Theodore A. Leisen 181

ZOÖLOGY

THE HABITAT OF THE BROOK TROUT IN MICHIGAN. T. L. Hankinson 197
SEASONAL VARIATION IN THE NUMBER OF VERTEBRAE OF FISHES. Carl L. Hubbs 207
ORTHOPTERA FROM GRATIOT COUNTY, MICHIGAN. H. M. MacCurdy 215
NOTES ON THE NESTING HABITS OF COTTUS. Bertram G. Smith .. 221

PLATES

PLATE FACING PAGE

I. *Halophila Engelmanii* Aschers. 10
II. *Halophila Baillonis* Aschers. 10
III. *Cymodocea isoetifolia* 10
Fig. 1. Male and Female Inflorescences of *Cymodocea isoetifolia* from Australia.
Fig. 2. Male and Female Flowers of *Cymodocea isoetifolia*.
IV. Leaf-blade, Shoot, Inflorescence of *Posidonia australis*... 10
V. *Armillaria dryina* 66
VI. *Armillaria macrospora* 66
VII. *Armillaria arenicola* 66
VIII. *Armillaria robusta* 66
IX. *Armillaria albolanaripes* 66
X. **Lipsanocystis traversensis**, n. sp.. 160
XI. Nesting Habits of Cottus 224

TEXT FIGURES

FIGURE PAGE

I. Drawing of *Thalassia testudinum* to show Habit, Flowers and Fruit 7
II. Drawings of Pollen Grains of *Thalassia testudinum*, showing the Germination of the Pollen Tubes 9
III. Graph representing the Change in Hydrogen Ion Concentration 51
IV–VI. Diagrams of Plates of **Lipsanocystis traversensis**....... 157
VII. Graph showing Water Temperatures of Douglas Lake in 1921 209

THE GENUS RIBES WITH REFERENCE TO THE WHITE PINE BLISTER RUST IN THE WEST

DOW VAWTER BAXTER

During the season of 1921, the Office of Blister Rust Control was obtaining information regarding the occurrence of pine and Ribes with reference to blister rust in the West. The area included in this paper embraces the Great Plains and the Rocky Mountain regions. It was essential to secure information regarding the kind of gooseberries and currants present, their abundance, and the continuity of the bushes along the streams crossing the great plains. In addition, it was necessary to determine whether or not they are found in the area of general distribution of native five-needled pine.

That Ribes may become one of the most important small fruits of this region is shown by the experiments of the various dry land experiment stations. At the Northern Great Plains Field Station, fourteen or fifteen of the principal varieties are under cultivation without irrigation, and the Belle Fourche experiments show that Ribes bear even better when grown on the dry land plots.

The most eastern outpost found for native *Pinus flexilis* was on Harney Peak (elevation 7240 feet), Black Hills, South Dakota. One hundred and eighty-six trees, including seedlings, were counted, averaging twenty-six feet apart or about sixty-five per acre. The trees were associated with *Pinus ponderosa.* Four specimens of *Pinus contorta* were also seen in this locality. Ribes were present; some bushes touched the small seedlings. They seemed, however, to grow more or less in spots; in some places they were not so common.

Pinus flexilis and Ribes (principally *R. cereum*) occur in the Laramie Mountains. This range, which lies between the great plains on the east and the Laramie Basin on the west, is a continuation of the front range of Colorado mountains. Here, on the west slope, plots which had different aspects were paced off near the summit. The pine and Ribes were counted on these plots in order to obtain a more accurate estimate of their relative abundance.

Near the summit of the range in a plot 250 feet square with a southern aspect, 18 specimens of *Pinus flexilis* were associated with 35 *Ribes cereum* bushes; the trees ranged from 1½ feet in height to 3 feet in diameter. The nearest bush to a limber pine was 8 feet away. On a plot 125 feet square having a western aspect, 25 limber pine were present, but no Ribes. On a similar plot 1½ miles from the first, but having an eastern aspect, 45 five-needled pine and 72 *Ribes cereum* bushes were found. Counts were made, likewise, on a strip about 35 feet wide for 1750 feet along a small running spring in Telephone Cañon. Here 67 Ribes were counted.

It was found that an unbroken chain of Ribes links the limber pines of the Laramie Mountains and the Medicine Bow ranges. In the latter of these mountains the Laramie River rises, crosses the basin and the Laramie Mountains to join the North Platte and the Missouri rivers. Along both of these rivers Ribes were found.

In many sections of Wyoming, Ribes are by no means common. This is especially the case of the country just west of the Big Horns. Although they were found along the streams on the plains east of this range, the streams were scarce and for the most part dry in August.

Such conditions prevail south of Buffalo in northwestern Wyoming where *Cronartium occidentale* was found on *Ribes aureum*. This find greatly extends the known distribution of this rust. As the identity of the Cronartium stage of the piñon blister rust and *Cronartium ribicola* is difficult to distinguish without inoculation experiments, field-work will be greatly complicated should the white pine blister rust continue to spread eastward from the northwest Pacific Coast.

University of Michigan

THE DISTRIBUTION AND POLLINATION OF CERTAIN SEA-GRASSES

H. H. M. BOWMAN

The sea-grasses are not really grasses, but are submerged, marine flowering plants belonging to two families, the Hydrocharitaceae and the Potamogetonaceae. These plants are little known and have been studied by few botanists. The sea-grasses are the only flowering plants which live in pure salt water and which can not exist in fresh or even brackish water. This limitation of the term "sea-grass" has been proposed by Dr. Ostenfeld of the Botanical Museum at Copenhagen (1). His definition excludes four related aquatic genera, viz., Ruppia, Zannichellia, Althenia, and Leptilaena, which live in fresh or brackish water and all of which are not entirely submerged.

The sea-grasses of the world are embraced in only eight genera, all belonging to the two families mentioned, and comprise collectively only thirty species. Of these eight genera, three are in the Hydrocharitaceae, namely, Halophila, Enhalus and Thalassia, and five are in the Potamogetonaceae, namely, Cymodocea, Diplanthera, Posidonia, Zostera and Phyllospadix. Of these the first two are closely related while Zostera and Phyllospadix are practically alike; some botanists recently have expressed doubts as to the standing of Phyllospadix. It is regarded by some as being merely a subgenus of Zostera. Zostera is perhaps the best known of all the sea-grasses, since it has migrated into the colder oceans and is widely distributed in its two species throughout the world. *Zostera marina* occurs all along the Atlantic coast from the Gulf of the St. Lawrence to Virginia.

The limited distribution of most of the sea-grasses is due to several factors. The most important restrictive factor is the metabolic one inherent in most of the group, i.e., the peculiar thermal and salinity requirements. Another limiting factor is the fact that these plants do not have special apparatus for dispersal by floating such as many marine or aquatic plants have. Even when tropical storms uproot them from the submerged meadows where several genera grow, the floating rhizomes die in a few days. It has been suggested that animals might be dispersive agents, such as fishes, turtles and marine mammals like the manatee and the dugong, but it has not been demonstrated that such is the case.

The light requirements, too, necessitate their growth in shallow water except in very clear oceans where the light can penetrate to greater depths, as I pointed out in a paper seven years ago (2). A consideration of all these distributional factors would indicate that the sea-grasses have a very slow rate of dispersal and are limited almost wholly to transportation by water currents carrying the seeds to suitable habitats.

In the waters surrounding this continent there are found only seven species of sea-grasses, and all of these except Zostera are limited to the Gulf of Mexico and the Caribbean Sea. The six tropical species are *Halophila Engelmannii, H. Baillonis, H. Aschersonii, Diplanthera Wrightii, Cymodocea manatorum* and *Thalassia testudinum*. Of these six species I have collected four, in the Gulf region, i.e., all except *Diplanthera Wrightii* and *Halophila Aschersonii*. It is probable that I have also collected the last named one, since it has the same range as *H. Engelmannii* and resembles it a great deal. Ostenfeld described this species in 1902, and in my paper referred to above, published in 1916, I mentioned the occurrence of only two species of Halophila in Florida waters and subsequently my friend, Dr. Barnhart of the Bronx Gardens, called my attention to Ostenfeld's more recently described species occurring in that range. In correspondence since with Ostenfeld he has agreed with me that the two species perhaps merge. The main specific difference is in the veination of the leaves.

Nearly all the sea-grasses are dioecious and it is only rarely that flowers of both sexes are found. Of course one reason for this is the difficulty encountered in collecting submerged marine plants. Small (3) in his *Flora of the Florida Keys*, published in

1913, says that the flowers and fruits of the American species of Halophila had not been found, but of the four species which I collected I was fortunate enough to get plants of at least one sex in three species and of both sexes in one. I attribute this to the method of collection. In 1915 and 1916 I made numerous trips about the Gulf of Mexico on the yacht of the Carnegie Institution, the *Anton Dohrn.* This vessel is equipped with deep-sea dredging and sounding apparatus and has a motor launch tender carrying air pumps and a diving helmet. The Halophila species I secured by using the dredge at a depth of sixteen to eighteen fathoms, that is, 96 to 108 feet. The turtle and manatee-grasses I collected by going down in the diving helmet in depths of twenty to thirty-five feet and digging out the rough rhizomes from the bottom. By the former method I secured *Halophila Engelmannii* (See Plate I) and *H. Baillonis* (See Plate II) with pistillate flowers and nearly ripe fruits, in June, about sixteen miles out in the Gulf stream south of the Tortugas Islands. The occurrence, however, of the sea-grasses in this region at such unusual depths is not without important significance, as has already been pointed out in my paper of 1916, and it is probable that these species can exist even here only at those depths on account of the great clarity of the water.

The distribution of the sea-grasses has by recent studies on their biology thrown light on the geographical and geological conditions of the past. Ascherson and Ostenfeld have shown that these lowly marine spermatophytes are very ancient genera and originated in the Indo-Pacific Ocean in early Tertiary time. Morphologically the Caribbean species are closely related to the species found at present in the Malay Archipelago and Australia. This indicates a common ancestry, but on account of the peculiar thermal and salinity requirements of these species, it is seen that their distribution helps to substantiate the theory that an ocean connection existed in former times between the Atlantic and Pacific oceans where the Isthmus of Panama now is. This conclusion has also been arrived at independently by other botanists from studies made on other plants, notably by N. Svedelius, who accords with George Murray's earlier theory (1873), from his studies made in 1906 on distribution of the marine alga, Caulerpa. The work of both these men as well as the studies of Ostenfeld give strong support to the hypothesis of a water passage between the two parts of the American continent. The distribution of the sea-grasses at the unusual depths at which I found them near the Tortugas Islands is also of geological interest. Their bathymetric distribution helps to substantiate the studies of Dr. Vaughan of the United States Geological Survey and his theory of the subsidence of the southern portion of the Florida peninsula and the Gulf floor in Pleistocene time, and in which Dr. Vaughan concurs, as pointed out in my earlier paper on the sea-grasses.

In addition to the pistillate plants of the two species of Halophila which I secured in the dredges, I also collected the female plants of *Cymodocea manatorum* inside the Tortugas Atoll by going down in the diving outfit and plucking it from the bottom. These plants are found growing in about twenty feet of water, forming immense green meadows or lawns on the ocean floor. This Cymodocea or manatee-grass is about six to eight inches tall and bears the flowers, which are very tiny, in the sheathing axils of the leaves. In all the collection over these submerged meadows only the pistillate plants were found. This, however, is not unusual since in a paper sent me recently from Copenhagen by Ostenfeld, describing a new species of Cymodocea which he collected in 1914 in Australia, while a guest of the British Association for the Advancement of Science, he is forced to say in the description, "*flos masculus ignotus.*" This difficulty is partly overcome by assuming that the male flower is like that of our Florida species, *C. manatorum,* since the female flowers of both the Australian and the Florida species are alike in so many respects. The illustration of the male and female inflorescences, Plate III, Fig. 1, which I present here are not those of *C. manatorum,* but of *Cymodocea isoetifolia,* a native of Australia, and which I have adapted from Ostenfeld to show the general habit of *C. manatorum.* In 1907 he and Warming collected *C. manatorum* plants in the Danish West Indies, now the

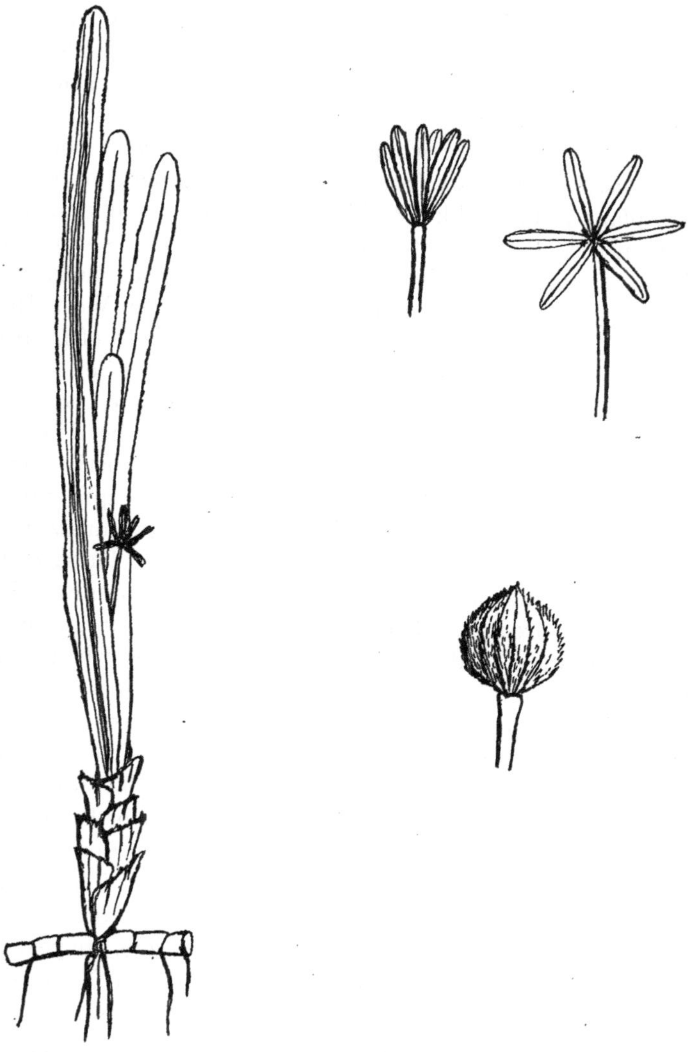

FIG. 1. Drawing of THALASSIA TESTUDINUM to show Habit, Flowers and Fruit.

American Virgin Islands, and describes at length the similarities of *C. manatorum* of America and *C. isoetifolia* (See Plate III, Fig. 2) of Australia (4). These species of Cymodocea have the same method of pollination that the other sea-grasses have.

The only sea-grass of which I collected flowers of both sexes was *Thalassia testudinum* (See Fig. 1). The general appearance of this plant is much like Posidonia, of which an illustration is also given (See Plate IV) to show the superficial resemblance in habit to Thalassia. This plant also forms great meadows in shallow bays and is abundant in nearly all the West Indies region as far north as Key West, occurring at a depth of from two to twenty feet. By wading out along the shore after a severe hurricane, when many plants had been torn up from the bottom by the large waves, I was able to collect from the débris numerous pistillate plants with fruits. The green bristly bur-like fruits are about one centimeter in diameter and split into six sections when ripe, setting free the seeds. The staminate plants have single flowers of six anthers borne on a stalk. These anthers are pinkish gray in color and were collected from staminate plants brought up in dredges. The figures of these two inflorescences of Thalassia were made from note-book sketches which I had hastily made.

The pollination of most all of the sea-grasses is greatly specialized and adapted to their submerged life. The process is in the main the same as that described by Clavaud (5) for *Zostera marina* which he observed along the coast of Brittany by gathering inflorescences and placing them in dishes of sea-water and observing the process with a microscope. The anthers of most of them are tetralocular and split extrorsely. The pollen is confervoid and the grains are of various shapes. In Halophila it is ellipsoidal, as in Zostera, but in Thalassia it is spherical. In this latter genus the pollen grains are packed in the anther sacs in coiled spiral masses and adhere to one another like beads in a necklace, i.e., the pollen is moniliform. When the anthers are ripe, the chains of pollen grains fall into the surrounding water; the gelatinous walls of the grains imbibe water and form a stringy mass of slippery mucus-like

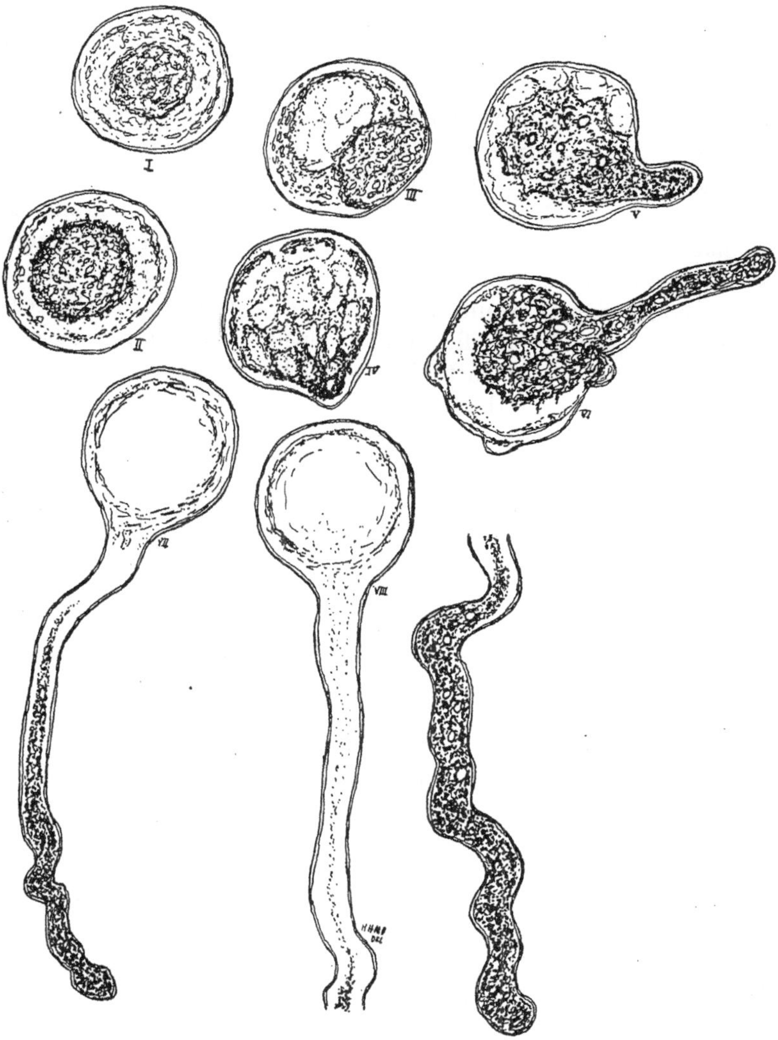

Fig. 2. Drawings of Pollen Grains of Thalassia testudinum, showing the Germination of the Pollen Tubes.

pollen. On taking some of this mucus-like material on a slide and examining it with a microscope, I found it to be a mass of pollen grains, many of which were germinating their pollen tubes. So far as I know, no one has ever seen the pollen tubes of any of the sea-grasses, or at least I can find no references or figures of them in botanical literature, and so I have presented here several note-book drawings of the pollen grains of *Thalassia testudinum* and of the germination of the tubes showing several stages in their development (See Fig. 2).

In conclusion, I would say that the distribution of the sea-grasses is especially significant in a consideration of the geology and geography of the past history of our American continents, more particularly in the taxonomic relations of the American and Australian species, as others have already pointed out, and also that the general morphology of a few typical genera of these little-known plants and description of their peculiar method of water pollination might not be uninteresting to botanists in general.

TOLEDO UNIVERSITY
TOLEDO, OHIO

REFERENCES

1. OSTENFELD, C. H. 1915. On the Geographical Distribution of the Sea-Grasses. Proc. Royal Soc. Victoria., N. S., 27. Part II.
2. BOWMAN, H. H. M. 1916. Adaptability of a Sea-Grass. Science, N. S., 43. No. 1103.
3. SMALL, J. K. 1913. Flora of the Florida Keys, New York.
4. OSTENFELD, C. H. 1916. Contributions to West Australian Botany, Dansk Botanisk Arkiv, Bind II. Nr. 6.
5. CLAVAUD, ARMAND. 1879. Sur le véritable Mode de Fécondation du Zostera marina. Ann. Linn. Soc., Bordeaux, p. 109.

PLATE I

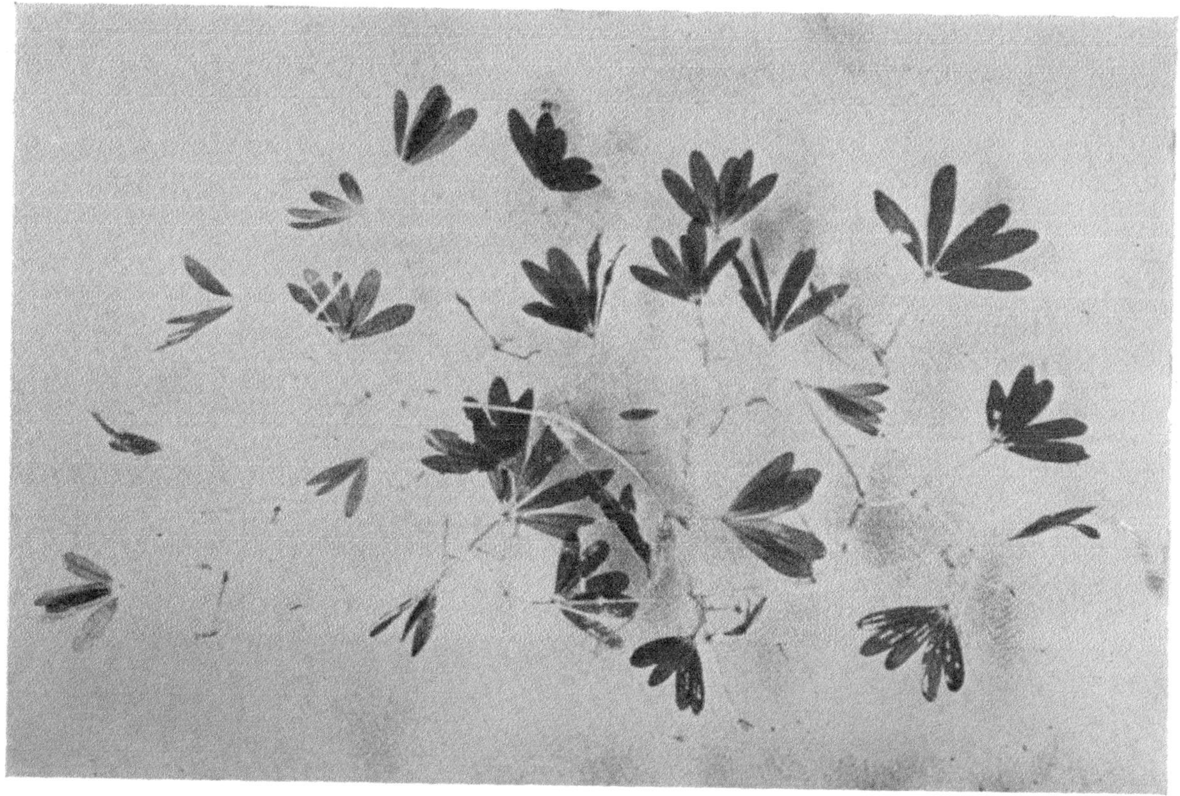

Halophila Engelmanii Aschers.

PLATE II

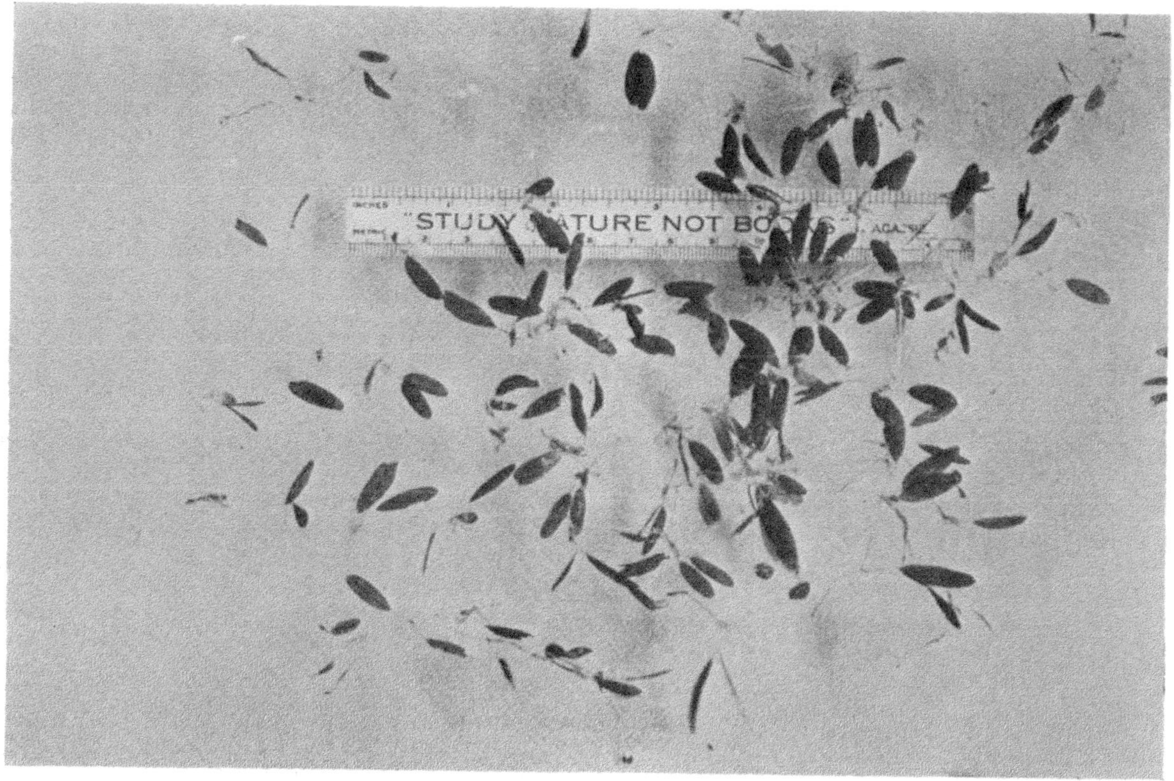

HALOPHILA BAILLONIS Aschers.

PLATE III

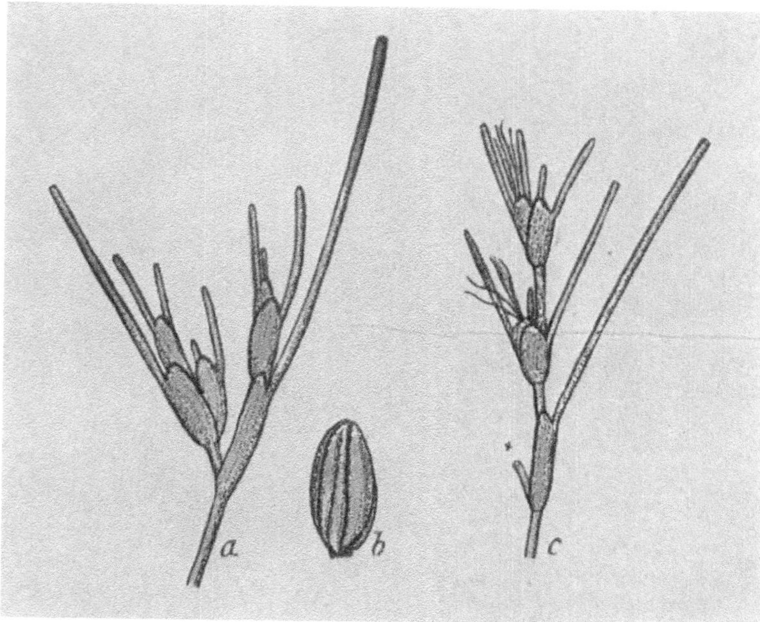

FIG. 1. Male and Female Inflorescences of CYMODOCEA ISOETIFOLIA from Australia (after Ostenfeld).

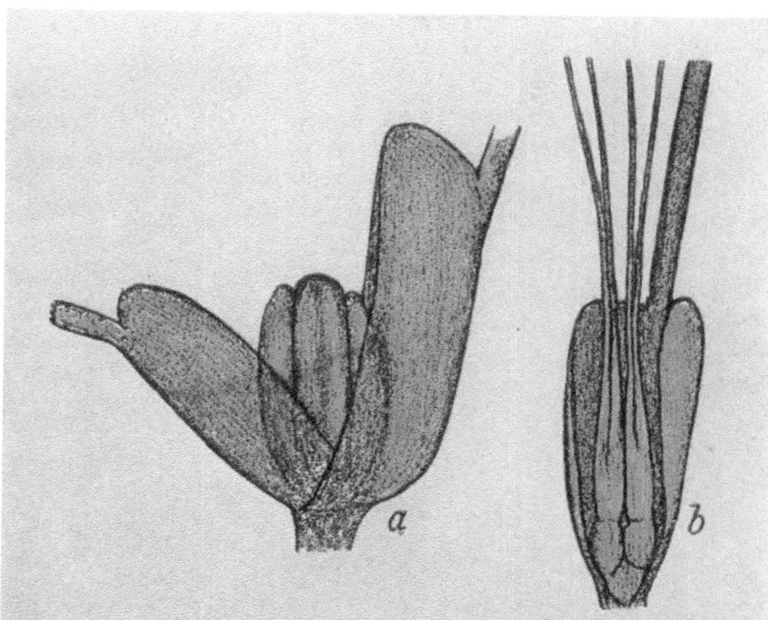

FIG. 2. Male and Female Flowers of CYMODOCEA ISOETIFOLIA (after Ostenfeld).

PLATE IV

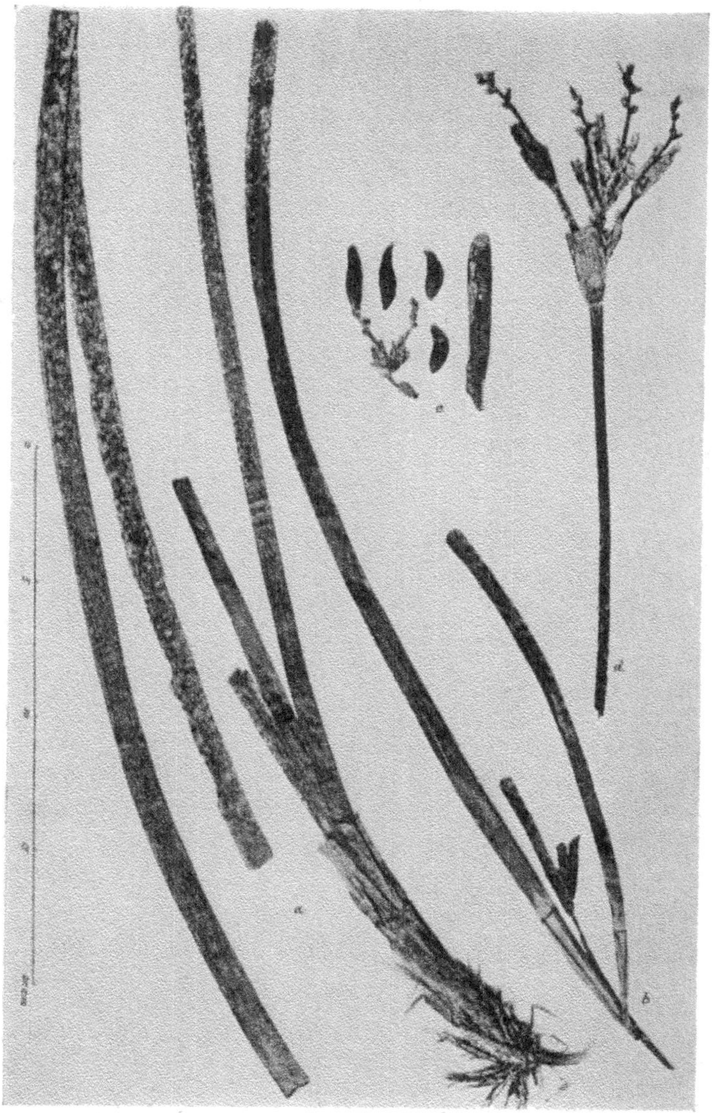

Leaf-blade, Shoot, Inflorescence of POSIDONIA AUSTRALIS, an Australian Species, showing the General Habit of the Genus (from photograph by Ostenfeld).

NOTES ON THE MICHIGAN FLORA

PART V*

OLIVER ATKINS FARWELL

The year 1921 was exceptionally auspicious for field work, being for the most part dry and hot. Fortunately what little rain there was, fell during the nights or on days not set apart for botanizing.

On April 8, I gathered a few specimens of *Carex varia,* the first time in my thirty years of collecting here that I have seen it within the city limits. On May 15, when Mr. Walpole of Ypsilanti took me around the country in the vicinity of that place and in Pittsfield Township, *Viburnum dentatum* was collected for the first time; it was in bud only, since the season was somewhat early for flowers. On the campus of the Normal School, *Galium verum* and *Thymus serpyllum* were growing in profusion, but there was no sign of flowers so early in the year. Mr. Walpole says that they have been there for many years. In company with Mr. Gladewitz, I found on June 9, *Thaspium barbinode* var. *angustifolium* growing in abundance a short distance north of Monroe; Mr. Billington found this in the southwestern part of the State. Near Washington on June 21, again in company with Mr. Gladewitz, I found *Solidago arguta* in flower. This is a very early record for the flowering of goldenrod in Michigan. On this day's tramp we found a grove of Ailanthus in all stages of development from the seedling to fruit-producing trees; there were hundreds of them covering a territory of several acres in extent. A peculiarity about these young trees of the open woods, not observed in those of the cities, is that the terminal shoots freeze back during the winter

* For Parts I, II, and III, see *Michigan Academy of Science, Annual Reports,* 20:161–195, 21:345–371, and 22:117–185, respectively; for Part IV, see *Papers,* 1:85–100.

and the trunks are continued by the development of branches from lateral buds.

While on my annual vacation on August 10, I saw for the first time at Lake Linden, the common wormwood, *Artemisia absinthium*, growing wild. On August 16, at Copper Harbor, a seedling of *Populus candicans*, only two feet in height, was observed on the banks of the harbor shore where there was no possibility of its having been placed by man's intervention. At the old Cliff Mine, my old home, on the same day, a few specimens of *Lycopodium inundatum* var. *Bigelovii* were seen. The next day at Calumet, *Matricaria suaveolens*, the anise chamomile, was found in profusion. *Mimulus moschatus* is frequent along the borders of streams in this region. Mr. Richard Ford of Ypsilanti collected it in August of 1921 at Munising, which is, so far as I am aware, the second locality reported for this plant in Michigan.

In 1909, the Detroit Mycological Club made a trip to Grosse Isle and Hickory Island, incidentally examining the work that was in course of progress on the Livingston channel. As a result of the coffer dam in the Detroit River, a large part of the river bed south of Stony Island was exposed to view. On this portion of the bed and on the southern half of Stony Island proper, *Salvia lanceaefolia* and *Euphorbia dentata* were found in great profusion. In August of 1921 another trip was made to Stony Island to look for Salvia, but not a vestige of either plant could be found. *Euphorbia dentata* was, however, found on the mainland at Trenton. Near Stony Creek, on a high bank made by cutting the road through a hill, *Schizonotus sorbifolius* is plentiful, now covering several times the amount of ground it did ten or fifteen years ago. On the bank across the road is a grove of a score or more of young Lombardy poplars, ranging up to twenty feet in height, which were not there a dozen years or so ago. Evidently they have not been planted by man. How they procured a footing in this out-of-the-way place is insolvable unless we assume that at some time many years ago a tree had been planted there, all evidence of which has now disappeared, and that the present grove is made up of suckers derived from the original tree. On September 28, near Farmington we found a small specimen of *Sambucus Canadensis* in flower, a thing that is very unusual at this time of the year. On a range of hills otherwise nearly bare we found not far from Goodison *Rosa rubiginosa* in greater abundance than anywhere else, counting more than fifty plants, each well laden with its large red hips. The more important collections, redeterminations, and critical notes for the year are listed below in systematic order.

THELYPTERIS DRYOPTERIS (L.) Slosson, THELYPTERIS PHEGOPTERIS (L.) Slosson, and **Thelypteris hexagonoptera** (Mx.) n. comb. The name Thelypteris Schmidel was applied by Dr. Nieuwland to Dryopteris Adans., but Miss Margaret Slosson interprets the name as more correctly applying to the genus known as Phegopteris Fée, with the inclusion of several of C. Christensen's subgenera of Dryopteris, basing the generic character chiefly upon the fact that these ferns have true hairs, and restricting Dryopteris to those species without true hairs. Our species would be *Thelypteris Dryopteris* (L.) Slosson, *Thelypteris Phegopteris* (L.) Slosson, and **Thelypteris hexagonoptera** (Mx.) n. comb. (*Polypodium hexagonopterum* Mx., *Fl. Bor. Am.*, II. 271. 1803). The latter species is occasional in the southeastern section of the State, but rare in the Upper Peninsula. Dearborn, August 15, 1920, No. 5604; Farmington, Sept. 28, 1921, No. 6069; Keweenaw Co., Sept. 20, 1888, No. 707; Geddes, August 21, 1909, No. 707a; Parkedale, August 4, 1912, No. 3035.

ATHYRIUM FILIX-FEMINA (L.) Roth, var. **commune** (Eaton) n. comb. (*Asplenium filix-femina* var. *commune* Eaton, *Ferns N. Amer.*, II. 227, Pl. 76, Fig. 5. 1880; *Athyrium angustum* var. *rubellum* Butters, *Rhodora*, 19:193. 1917, in large part.) This, the commonest form of the species in this country, has twice pinnate fronds and serrate to pinnatifid pinnules. As pointed out by Butters, the indusia of this form are not glandular-ciliate, as in the typical European form, but the two forms are connected through the variety *Asplenioides* (*Athyrium filix-femina* var. **Asplenioides** (Michx.) n. comb., *Nephrodium asplenioides* Michx., *Fl. Bor. Am.*, II. 268. 1803), which, like the

European form, has glandular-ciliate indusia. Detroit, Oct. 16, 1910, No. 502a; Utica, Oct. 5, 1921, No. 6079; Oxford, Oct. 12, 1921, Nos. 6113, 6119, 6120; Farmington, Sept, 28, 1921, No. 6065. There is a color variation of this which has the stipes, rachis and rachiola red and may be known as forma **rubellum** (Gilbert), n. f. (*Asplenium filix-femina* D. C. Eaton, *Ferns N. Am.*, II, 228–230, Pl. 76, Fig. 1. 1880; *Athyrium filix-femina* var. *rubellum* Gilbert, *List N. A. Pterid.*, 35. 1901). Rochester, Oct. 19, 1921, No. 6125.

FILIX THELYPTERIS (L.) Farwell. A common fern in swamps. The typical form has the frond as much as 90 cm. in length; the stipe is a little shorter or a little longer than the blade, which is as much as 15 cm. in width; the lobes of the pinnae are oblong, the sterile ones broadly so, 6 or 7 mm. long, and obtuse; the fertile frond usually is smaller and more coriaceous than the sterile; the pinnules become revolute and acute. Utica, Oct. 5, 1921, No. 6085; Goodison, Oct. 7, 1917, No. 4690; Oxford, Oct. 22, 1920, No. 5615; Detroit, Sept. 30, 1892, No. 1348; Rochester, Aug. 15, 1909, No. 1348a; Parkedale, Aug. 11, 1912, No. 3038a; Algonac, Sept. 12, 1914, No. 3885. In a tamarack swamp near Oxford there was found a much larger form with fronds often 1.3 m. in height and 25 cm. broad; the pinnules from 8–14 mm. long, linear-oblong, normally obtuse or acute, the fertile often sharply acute from the strongly revolute margins. It may be known as var. **linearis,** n. var. Oct. 10, 1921, No. 6114. Some clumps of this variety have no normally coriaceous, fertile fronds, but some of the delicate sterile fronds have more or less fructifications on the unchanged pinnules on the upper half of the blades; the sori are not confluent. This may be known as forma **frondosa,** n. f. Oct. 10, 1921, No. 6115.

FILIX BOOTTII (Tuckerm) Farwell. One of the rare ferns in Michigan. New localities: Oxford, Aug. 22, 1920, No. 5615f; Franklin, June 24, 1920, No. 5505; Utica, Oct. 5, 1921, No. 6087a.

Felix Dowellii, n. sp. (*Dryopteris Clintoniana* × *intermedia* Dowell, *Bull. Torr. Club*, 35:136. 1908.) Utica, June 6, 1916, No. 4179a.

FILIX BENEDICTII, n. sp. (*Dryopteris Clintoniana* × *spinulosa* Benedict, *Bull. Torr. Club*, 36:45. 1909.) Rochester, Oct. 6, 1918, No. 5186.

FILIX SPINULOSA (L.) Farwell. Sometimes as much as a metre or more in height with the stipe somewhat shorter than the blade; lowest pinnae 20 or more cm. in length by 14 or 15 in width at the base; lowest anterior pinnule 10 cm. long and varying from 1.25 to 2 times the length of the lowest posterior pinnule. Usually the fronds of this species are much smaller, about 1/3 to 2/3 the size given above; the stipes are variable in length, but usually those from the same crown are of about the same length; some of the fronds are erect, others spreading. The variety *intermedia* is variable in much the same way, but it is darker green; the pinnae are closer or overlapping, and the second and often the third anterior pinnules are longer than the first. Both are common in low, wet, swampy or alluvial woods.

LYCOPODIUM INUNDATUM L. var. BIGELOVII Tuckerm. This variety is said to be taller and more slender than the species with the leaves of the spike incurved or appressed, occurring mainly near the coast in New England and southward to Maryland. In Keweenaw Co., a few plants were found that were more slender than usual, but not taller, with more or less incurved sporophylls which very probably belong here rather than with the typical form of the species. August 16, 1921, No. 5971½.

TYPHA LATIFOLIA L. South Rockwood, June 14, 1921, No. 5868. Var. *remotiuscula* Simonkai. South Rockwood, June 14, 1921, No. 5869. The variety differs from the species in having the staminate spike separated from the pistillate.

HOLCUS SORGHUM L. var. SUDANENSIS (Piper) Hitchc. The cultivation of Sudan grass is of recent introduction and it may become adventitious in places. Shelbyville along roadsides, July 31, 1921, No. 5955a. Occasional; whether it becomes a permanent fixture remains to be seen.

MUHLENBERGIA AMBIGUA Torr. var. FILIFORMIS (Muhl.) Farwell. (*M. foliosa*, of Gray's *New Manual* as to the awnless form.) This species is not credited to Michigan in the *Michigan*

Flora, though it is frequent in the southeastern section and possibly in other parts also; perhaps it is included in *M. Mexicana,* which is there said to be variable. It prefers swampy grounds in the open or along the edges of sparsely wooded lands and very closely resembles the *M. diffusa* (*M. sylvatica* Torr.) of wet woods, but can be readily differentiated by the heavy glomerate, purple spikelets. Both have long exserted panicles; the spikelets of *M. diffusa* are green, slender and not at all glomerated. Bloomfield, August 22, 1916, No. 4395a, and Sept. 16, 1917, No. 4606a; Franklin, Sept. 23, 1918, No. 5170; Goodison, Sept. 4, 1919, No. 5382; Detroit, Aug. 29, 1920, No. 5623; Washington, Sept. 23, 1921, No. 5994; Ypsilanti, Aug. 31, 1919, No. 5361.

Danthonia compressa Austin. Much like *D. spicata* for which, without a close examination, it will generally be mistaken, but the long teeth 2–3 mm. of the lemmas, the terminal third of which are distinct awns, prove it to belong here. Washington, June 21, 1921, No. 5892.

Eragrostis pectinacea (Mx.) Steud. var. spectabilis (Pursh) A. Gray. A form of the species has glabrous sheaths; those of the specific type are pilose. Stony Creek, Gladewitz, Sept. 14, 1921; Farwell, No. 6010.

Bromus inermis Leyss. var. pellitus, Beck. The typical form of this species is glabrous throughout. It was reported a year ago from Algonac. The variety has the leaves and their sheaths more or less hirsute, the lower ones strongly so. Washington, Gladewitz and Farwell, June 21, 1921, No. 5893.

Stenophyllus capillaris (L.) Britt. var. cryptostachys Fernald. This variety is characterized by its longer, more numerously flowered, sessile spikes, rarely with one or two spikes short-pedicelled; the basal inflorescences, when present, are sessile. Utica, Farwell and Gladewitz, Oct. 5, 1921, No. 6096. Plants collected at the Zoölogical Park near Royal Oak by Billington and near Ypsilanti by Billington, Farwell and Walpole, previously reported as *Stenophyllus capillaris,* belong to this variety.

Carex cristata Schw. There are three distinct forms of

the inflorescence of this species which correspond to those of *C. scoparia*, as given in Gray's *New Manual.* The typical form has a linear-cylindrical inflorescence as much as 4 cm. in length; the spikes are distinct, but not separated. The var. *ellipsoidalis* has a congested, ellipsoid or ovoid head 2 or 3 cm. in length; the spikes are not distinct. Another form has moniliform flexuous inflorescence often 5 or 6 cm. in length and may be known as var. **catelliformis**, n. var. South Rockwood, Farwell and Gladewitz, July 12, 1921, No. 5928.

Carex festucacea Schk. not of American authors. Mr. Mackenzie has pointed out that the real *C. straminea* Willd. has *obovate* perigynia and is closely related to *C. alata*, separating from it *C. tenera* Dew. and *C. festucacea* Schk. The typical form has a moniliform inflorescence, spikes with clavate bases, perigynia about 3.5 mm. in length, with the conspicuous beak about as long as the orbicular body which is about 1.5 mm. broad. Trenton, May 26, 1921, No. 5772; Dearborn, May 30, 1921, No. 5795a.

Var. *tenera* (Dew.) Carey, has the spikes rounded at base, perigynia about 3 mm. long with the inconspicuous beak less than .5 as long as the orbicular body, which is nearly or quite 2 mm. broad. This is the commonest form of the species here. Keweenaw Co., Sept. 15, 1889, No. 683b; Port Huron, June 23, 1918, No. 4975a; South Rockwood, June 14, 1921, No. 5862; Detroit, July 21, 1892, No. 683a; Trenton, May 26, 1921, No. 5780; Washington, June 21, 1921, No. 5889.

A similar variety with somewhat longer perigynia, body more gradually tapering into the beak, which is spreading or conspicuous, is **C. festucacea** var. **echinodes** (Fernald), n. comb. (*C. straminea* var. *echinodes* Fernald, *Proc. Am. Acad. A. and Sci.*, 37:474, Fig. 30. 1902.) Rockwood, June 16, 1918, No. 4948; Washington, June 21, 1921, No. 5891; Keweenaw Co., June 19, 1903, No. 1801. Another variety with larger, ferruginous ovate spikes, round at base is var. **ferruginea** A. Gray, n. comb. *Carex foenea* var. *ferruginea* A. Gray, *Man.*, 580. 1867). Rarely met with. Detroit, June 7, 1895, No. 970; Washington, June 21, 1921, No. 5898.

Carex stramimea Willd. Very much like a very slender *C. alata* Torr., with which it is probably conspecific. Spikes clavate-based, perigynia 3.5 mm. long, body obovate, 2.25 mm. wide, beak about one-half the length of body or less. Woodville, August 4, 1921, No. 5958. A tall robust variety with perigynia 6–7 mm. long and 3.5 mm. wide is var. *alata* (Torr.) Bailey. Keweenaw Co., July 18, 1890, No. 762; Royal Oak, Billington and Farwell, August 9, 1920, No. 5578.

Carex brevior (Dew.) K. K. Mack. Usually about .5 m. in height, but occasionally as much as 1.25 m.; inflorescence of 3–6 approximate spikes. Detroit, June 15, 1893, No. 1369; Dundee, July 15, 1918, No. 5294; Rockwood, June 16, 1918, No. 4940; Harris, July 13, 1918, No. 5081; La Salle, June 2, 1921, No. 5817; South Rockwood, June 14, 1921, No. 5861 and July 12, 1921, No. 5919 (coarse plants 1.25 m. in height with pathological perigynia). Stony Creek, June 9, 1921, No. 5823. Another form of this species usually a meter or more in height has 6–10 distinct spikes and may be known as var. **pseudofestucacea**, n. var. (*Carex festucacea* Fernald, *l. c.*, 477, Figs. 47–48, not of Schkuhr.). Keweenaw Co., August, 1901, No. 1789.

Carex conjuncta Boott. One locality, Manistee, is given in the *Michigan Flora* and one, Keeler, in the *Supplement.* I collected it at Detroit, June 30, 1907, No. 2030.

Carex rosea Schk. A coarse, often rigid plant, with numerously flowered spikes and leaves 2–3.5 mm. wide, often with the aspect of *C. echinata* var. *cephalantha.* Parkedale, June 11, 1912, No. 2679 and August 9, 1914, No. 3819; Detroit, June 8, 1893, No. 1361; Washington, June 21, 1921, No. 5895; Dearborn, May 30, 1921, No. 5796. A more common form is var. *minor* Boott., which is a lower plant, quite slender, with leaves about 1.5 mm. wide, fewer flowered, more distant spikes, and smaller perigynia which are *not* ascending, as described in Gray's *New Manual;* at least mature perigynia are not, as it is found here; stems erect. Usually in open copses on dry soil. Trenton, May 26, 1921, No. 5782; Detroit, June 16, 1907, No. 2023. A form of this variety is found in the shade of rich, moist woods, with much longer, prostrate stems. It has passed as var. *radiata*, but can not be Dewey's variety of that name, which is a plant with setaceous stems and leaves with the aspect of *C. trisperma* and not found as far west as Michigan. It may be known as *Carex rosea* var. *minor* forma **debilis,** n. f. Trenton, May 26, 1921, No. 5781; Detroit, June 24, 1893, No. 1380.

CAREX RIPARIA var. IMPRESSA S. H. Wright. A constant search for this variety extending over many years has finally resulted in success. It was found in low, wet grounds, in a sparse and narrow copse near Trenton. The lower sheaths are not purple and all bear long leaves longer than the culm, which gives the plant quite a different aspect from the ordinary form with leafless, purple sheaths. Sheaths usually not fibrillose; only one plant was detected with one or two sheaths fibrillose and these were situated about one-third up the culm. Nerves impressed and very inconspicuous. Plants about two feet, much shorter than the common form of the species. May 26, 1921, No. 5788.

ARISAEMA TRIPHYLLUM (L.) Torr. var. **viride** (Engler) n. comb. (*Arisaema atrorubens* var. *viride*, Engler in *DC. Mon. Phar.*, II. 536. 1879.) Spathe wholly green. Pittsfield, Walpole, Ford and Farwell, May 15, 1921, No. 5734.

TRADESCANTIA REFLEXA Raf. Only three or four localities for this species are given in the *Michigan Flora.* It seems to be pretty well distributed in the southeastern section of the State though not frequent at any one place. Monroe, June 9, 1921, No. 5831; River Rouge, Sept. 15, 1918, No. 5141; Detroit, July 12, 1905, No. 1915.

LETHEA, Norrohna, Verh. *Batav. Gen. V.*, Art. IV, 2. 1790. (*Disporum* Salisb., *Trans. Hort. Soc.*, 1:331. 1812.)

Lethea is said to be a *nomen nudum* and therefore not available, but many maintain that any name, the identity and application of which are well known and not questioned, is properly published. Lethea will, therefore, be used here in place of Disporum.

Lethea Cahnae, n. sp. Stem up to the first branch about 3 dm. in height at flowering time, clothed with a fine short pubescence, with 3 or 4 sheathing scales near the base and one about 1 dm. below the branch, similarly pubescent and about 4 cm. long; branches similarly pubescent, often naked on the lower half, foliaceous on the upper half and terminated by one or two flowers; leaves sessile, finely pubescent on both surfaces, oval or oblong, short acuminate, rounded at base, about 5–9 cm. in length by 2.5–4 cm. in width, prominently 5–7 nerved with intermediate, less conspicuous nerves; flowers 3–4 cm. across when open, often 3 cm. to tip of style, which is longer than the equal stamens and perianth segments; segments ovate-lanceolate or, to be more accurate, trowel-shaped, narrowed to distinct greenish white claws about 5 mm. long; blade from 15–20 mm. long by 7 or 8 mm. wide near the base; glabrous, white, copiously spotted on the inner surface with small dots of a madder purple color, filaments slender, slightly flattened 14–17 mm. long, anthers 5–7 mm. long; style 20–23 mm. long, lobes of the stigma 3, about 1 mm. long; ovary 2–3 mm. long, deeply 3-lobed, densely pubescent, sessile; fruit unknown, but a plant collected by Mrs. Cahn in 1921, after the flowers had fallen, gave evidence that the ovary developed a short stipe; pedicles about 1 cm. long, finely pubescent; pubescence extending upward on the under surface of the lower part of the claws of the perianth segments. Related on the one hand to **Lethea Oregana** (S. Wats.) n. comb. (*Prosartes Oregana* S. Wats., *Proc. Am. Acad.* 14. 271. 1879) in its pubescent, obovoid ovary, and on the other hand to **Lethea trachycarpa** (S. Wats.) n. comb. (*Prosartes trachycarpa* S. Wats., *Bot. King. Explor.*, 344. 1871) in its three-lobed stigmas and in its leaves, but differing from both in its perianth characters and in the widely open flowers. May, 1921 and 1922. Near Farmington Center, Oakland County, where it was discovered by Mrs. Cahn of Highland Park, for whom it is named. The stamens are said to be hypogynous in this genus, but in this species they are attached to the bases of the claws and fall away with the segments.

Polygonum aviculare L. var. vegetum Ledeb. On sparsely wooded and pastured lands at South Rockwood, Gladewitz and Farwell, July 12, 1921, No. 5926.

Polygonum Pennsylvanicum L. The common form of this species in Michigan is var. *laevigatum* Fernald. Ever since this variety was published, a diligent search for the type has been kept up and this effort was rewarded on July 26, 1921, with its discovery at Trenton where it was found along the edges of a cultivated field. As there were only a few plants in that station, it may be that their presence there was the result of importation in grain seed; on the other hand, the station is within its natural range of distribution and the few plants seen may be the remnants that have escaped the natural processes of eradication. No. 5934; also found at La Salle, Sept. 21, 1921, No. 6039½ on low, alluvial grounds in company with the var. *laevigatum*, No. 6036. The two forms insensibly pass one into the other. On Slocum's Island there was found a form of the var. *laevigatum* with white flowers and this may be known as forma **albineum**, n. f. Farwell, Ford, and Walpole, August 31, 1921, No. 5980.

Polygonum arifolium L. In the *Michigan Flora* it is said to be infrequent, a statement which is probably quite accurate as it was collected this year for the first time in Michigan in nearly forty years of field work. In a wooded swamp near Oxford, Gladewitz and Farwell, Oct. 12, 1921, No. 6107.

Amaranthus Torreyi (A. Gr.) Benth. In September, 1918, a plant found in the railroad yards of Detroit was collected and labeled *Acnida sp.*, with a question mark. It was laid aside for future study and then forgotten. When it was taken up again recently, it appeared upon closer study to be the staminate plant of some dioecious Amaranthus, possibly *A. Torreyi*. A specimen was sent to Dr. Paul C. Standley, in whose absence Mr. S. F. Blake confirmed my identification as herewith given. At the time of collection the plant was badly injured and insect-eaten, showing only small leaves of late seasonal growth. Only the staminate plant was seen, No. 5149.

Allionia aggregata var. hirsuta (Pursh) Farwell. The form has the flowers in a terminal panicle. Rochester, Gladewitz and Farwell, July 21, 1921, No. 5874. Both the type and the variety at Ypsilanti and Rochester were found in the railway yards of those places, indicating that these forms may be introductions.

Scleranthus annuus L. A weed in pasture lands near Portage Lake; Walpole, June 11, 1921; Farwell, Billington and Gladewitz, June 12, 1921, No. 5855.

Silene antirrhina L. var. divaricata Robinson. A slender, woods form with pedicels divaricate and apetalous flowers. Portage Lake, June 12, 1921, No. 5840.

Ranunculus trifoliatus Muhl. This is one of the many forms that have gone to make up the polymorphous *R. septentrionalis* of our local manuals. An erect, rarely stoloniferous plant of dry grounds, early becoming glabrous. Petals roundish obovate 9-10 mm. in length. Stony Creek, Farwell and Gladewitz, June 9, 1921, No. 5819; La Salle, May 19, 1921, No. 5763.

Ranunculus caricetorum Greene. Another segregate of *R. septentrionalis*, prostrate, with the larger leaves usually biternate, and deltoid in general outline. La Salle, Farwell and Gladewitz, June 2, 1921, No. 5816; Trenton, May 26, 1921. No. 5776. Petals 10 or 11 mm. long, round obovate. Greene says it grows only in wet meadows where Caltha and Carex grow. The Trenton plant, apparently the same, was found in low depressions of pasture land retaining moisture for a long period of time.

Ranunculus septentrionalis Poir. Copiously hirsute throughout, erect, becoming prostrate and stoloniferous, leaves pinnately 3–5 foliolate, petals round-obovate, 8–13 mm. in length. Usually found in open woods and fields. Pittsfield, May 15, 1921, Nos. 5735, 5738; Detroit, June 17, 1893, No. 1129a; Dearborn, May 19, 1917. A prostrate stoloniferous form from swampy woods, usually smooth, and with trifoliolate leaves, is var. *nitidus*, Chapm. Dearborn; May 30, 1921, No. 5802. Perhaps this variety belongs more appropriately to *R. trifoliatus*.

Ranunculus lucidus Poir. (?) An erect coarse plant with stems 8 to 10 mm. in diameter, twice as large as in any of the preceding plants; glabrous or sparsely hirsute below; whole plant shining; branches elongating and rooting at the joints; leaves trifoliolate, often 20 cm. long and broad; divisions irregularly 2 or 3 parted, oblong or obovate, incised and serrate; flowers larger than in any of the preceding, petals round-obovate, 14–16 mm. in length. Alluvial grounds. La Salle, Farwell and Gladewitz, May 19, 1921, No. 5762, and June 2, 1921, No. 5815.

RANUNCULUS HISPIDUS Mx. Erect, 60 cm. in height, hirsute with retrorse hairs; leaves pinnately 3–5 foliolate, rather small, 5 to 10 cm. long and broad petals oblong-obovate, 6–8 mm. in length; roots fleshy, thickened. Ypsilanti, May 15, 1921, No. 5737; Detroit, May 19, 1907, No. 1905a.

SYNDESMON THALICTROIDES (L.) Hoffing. var. **biternatum,** n. var. Involucre apparently of 3 petioled ternate leaves, but actually of 3 sessile biternate leaves. Southfield, May 6, 1917, No. 4422; Geddes, August 21, 1899, No. 131b; Keweenaw Co., July 10, 1884, No. 131; Detroit, May 1, 1896, No. 131a. There is a similar plant that has the involucre consisting of only 2 leaves and this may be known as forma **oppositifolium**, n. f. Redford, May 24, 1918, No. 4858. In the preceding plants the flowers are umbellate; but there is another form with but one flower; otherwise like forma *oppositifolium* and this may be known as forma **unipediculum**, n. f. Redford, May 24, 1918, No. 4864½.[1]

LIRIODENDRON TULIPIFERA L. In the *Michigan Flora* this is described as occurring in the southwest section of the State only, e.g., at Ionia, Saranac, Lansing and southward. It occurs in Wayne, Oakland and Macomb counties in the southeastern section. Detroit, Oct. 25, 1895, No. 1665; Rochester, August 29, 1909, No. 1665a; Cass Lake, May 25, 1899, No. 1665b; Washington, Gladewitz and Farwell, September 14, 1920, No. 6003.

BERTEROA INCANA (L.) DC. One of the "alyssums" which is spreading as a weed in waste places. Rochester, July 4, 1910,

[1] All specimens of this species east of the Appalachian Mountains which I have seen have an involucre of sessile ternate leaves appearing as simple petiolate leaves and this form is the type of the species; of this Pursh described a one-flowered variety which may be known as *Syndesmon Thalictroides* forma **uniflorum** (Pursh), n. comb. (*Anemone thalictroides* var. *uniflora* Pursh. *Fl. Sept. Am.*, 11.387. 1814.)

No. 2170; Bloomfield, June 15, 1916, No. 4188; Galesburg, July 20, 1920, No. 5058; Calumet, August 17, 1921, No. 5972.

Barbarea verna (Mill.) Asch. Another locality was found for this species during 1921. This was along the railroad tracks near Erie in Monroe Co., where several large clumps of it were growing luxuriantly. May 10, 1921, No. 5729.

Cheirinia repanda (L.) Link. This species is not listed in the *Michigan Flora,* and is mentioned only in the *Synoptical Flora* as having been collected on made land around New York City and Philadelphia. Britton and Brown give it an additional range from Ohio to Kansas and Arizona. Farwell and Gladewitz found it in great profusion in an old garden near La Salle, May 19, 1921, No. 5757. Apparently it gives great promise of becoming as pernicious a weed as the charlock. In the *Synoptical Flora* it is mentioned under Erysimum, Tourn.; but Tournefort's Erysimum is typified by the hedge mustard, and the American species usually referred to it should be placed under Cheirinia Link, as is done by Britton and Brown.

Another species found in the Upper Peninsula is **Cheirinia parviflora** (Nutt.) n. comb. (*Erysimum parviflorum* Nutt. *ex* T. & G., *Fl. N. A.*, I. 95. 1838; *E. syrticolum* Sheldon, *Bull. Torr. Club*, 20:285. 1893.) I have seen collections from Colorado, Montana, Minnesota and Michigan and can find no grounds for separating the plants of Minnesota and Michigan from those of the more western states. It is simply that the species has emigrated eastward along the railways, an occurrence that is common enough in the history of plants. Keweenaw Co., June 26, 1895, No. 844.

Heuchera hirsuticaulis (Wheelock) Rydb. Another station was found for this species on the south bank of the Otter River near La Salle. Farwell and Gladewitz, June 2, 1921, No. 5766.

Schizonotus sorbifolius (L.) Lindley. Near Stony Creek on high banks along the roadside, Sept. 14, 1921, No. 6012. Collected here some twenty-five years ago and then reported as Spiraea and again later as Sorbaria; it appears that the earliest and valid generic name is Schizonotus of Lindley. Specimens are from plants that were originally cultivated nearby, but all traces of the residence have disappeared many years since.

Geum vernum (Raf.) T. & G. Two stations are given in the *Michigan Flora* and a third was reported by myself in the *Flora of Parkedale Farm*. A second station has been found in Wayne Co. Trenton, Farwell and Gladewitz, May 26, 1921, No. 5783.

Rosa acicularis Lindley. Listed in the *Michigan Flora* as *R. Engelmanni*. Keweenaw Co., Aug. 22, 1884, No. 185; Aug. 1, 1890, No. 779 and Aug. 18, 1890, No. 779a; Aug. 16, 1921, No. 5970. It has obovate fruit with a distinct neck. A form of this species has globular fruit and has been named var. *Bourgeauiana* Crepin. This variety is listed in the *Michigan Flora* as *R. Sayi* Schw. Keweenaw Co., August 22, 1884, No. 187. The leaflets in both forms are pubescent and resinous underneath.

Rosa spinosissima L. The Scotch rose is an escape from cultivation. Keweenaw Co. July 3, 1915, No. 3996; Trenton, Gladewitz and Farwell, May 26, 1921, No. 5793.

Rosa rubiginosa L. The sweetbrier is a common rose along roadsides and in fields throughout the southeastern section of the State. Near Goodison, Gladewitz and Farwell, Oct. 20, 1921, No. 6129; Ypsilanti, June 25, 1892, No. 1248. The scentless and glandless variety, *micrantha* (Sm.) Lindl., is very rare. Slocum's Island, Farwell, Ford, Gladewitz and Walpole, Aug. 31, 1921, No. 5981. In the specific type the fruit is obovate; in the variety usually flask-shaped, but often with some obovate fruits also.

Rosa blanda Ait. The lower part of the stem usually is copiously bristly; all the upper parts are glabrous and generally glaucous. In low usually moist grounds. Rochester, June 16, 1909, No. 186a; June 11, 1912, No. 2663; June 30, 1918, No. 5004; Aug. 15, 1909, No. 695b; Detroit, June 20, 1893, No. 695a; Keweenaw Co., Sept. 10, 1888, No. 695. A form has all stems and branches, both young and old, thinly but uniformly bristly, and the pedicels, ovary and fruit, are sometimes glabrous and sometimes glandular-bristly. It may be known as var. **hispida**, n. var. It agrees very well with the description of

Rosa blanda × *Virginiana*, but the leaflets are like those of normal *blanda* and not shiny above. Belle Isle, June 4, 1895, No. 971½. A low form with upper parts glabrous, purplish and glaucous, with the stipules, petioles, rachis, and midveins underneath quite glandular, and with the fruit globular to elliptic and with a distinct neck is var. *glandulosa* Schuette; common in dry, rocky fields, Keweenaw Co., Aug. 22, 1912, No. 3065.

Recently the name *Rosa Carolina* L. was transferred from our common swamp rose to our common pasture rose and in the *21st Annual Report*, pages 366–367, July, 1920, I followed this lead. *R. Carolina* L. *Sp. Pl.*, I. 492 (1753), was founded exclusively upon Dill., *Elth.* 325, t. 245, f. 316. This evidently is our swamp rose; the leaflets are too narrow for those of our field rose. The oldest name for our pasture rose is *Rosa Virginiana* Mill. (*R. lucida* Erhr.) *R. obovata* Raf. becomes *R. Virginia* var. **grandiflora** (Baker) n. comb. *R. humilis* Marsh. becomes *R. Virginiana* var. *humilis* (Marsh.) C. K. Schneid. *R. serrulata* Raf. becomes **R. Virginiana** var. **glandulosa** (Crepin) n. comb. These names supersede those listed under *R. Carolina* in the *21st Annual Report, l. c.*

CRATAEGUS OXYACANTHA L. var. MONOGYNA (Jacq.) Saniv. An occasional escape. Reported some years ago as *C. Oxyacantha* L. Detroit, June 9, 1901, No. 1715. Also on the hillside between the road and the shore at Grosse Isle, Farwell, Ford, Gladewitz and Walpole, Aug. 31, 1921, No. 5990. The nut has one seed in the variety while in the typical form of the species it is two-celled and two-seeded.

CRACCA VIRGINIANA L. Rochester, July 13, 1917, No. 4521; Portage Lake, Billington, Farwell, Gladewitz and Walpole, June 12, 1921, No. 5835; Island Lake, July 16, 1905, No. 936a; Orion, Aug. 29, 1895, No. 936. Rare; usually on light, sandy soil.

ROBINIA PSEUDO-ACACIA L. Near Goodison, Oakland Co., in a dense copse of native shrubbery and trees. Certainly not a cultivated tree in that locality. Farwell and Walpole, Sept. 14, 1919, No. 5383; Geddes, Aug. 21, 1909, No. 1133a.

ROBINIA VISCOSA Vent. Near Ypsilanti, June 6, 1891, No.

1133; Detroit, June 9, 1899, No. 1630. Similar to the preceding species, but the branchlets are clammy and the pods glandular-hispid.

MEIBOMIA CANESCENS (L.) O. K. One of our rare species of tick-trefoil. La Salle, Farwell and Gladewitz, Sept. 21, 1921, No. 5814; Grosse Isle, Aug. 14, 1909, No. 2101.

MEIBOMIA ILLINOENSIS (A. Gray) O. K. Several large clumps of this species were found near Portage Lake, but at that time it was too early for flowers. Billington, Farwell, Gladewitz and Walpole, June 12, 1921, No. 5846.

LESPEDEZA NUTTALLII Darl. Listed in the *Michigan Flora* on the authority of Britton and Brown. It looks a great deal like *L. frutescens*, but the stems are villous and the peduncles are longer than the leaves. Near Oxford, Farwell and Gladewitz, Oct. 12, 1921, No. 6108. All fruits had fallen, also the leaves of the stems, but the branches still bore their abundant foliage.

VICIA ANGUSTIFOLIA L. This was found growing in great abundance not far from Monroe in fields near the electric railway lines. Farwell and Gladewitz, July 6, 1921, No. 5912.

LATHYRUS VENOSUS Muhl. var. INTONSUS Butters & St. Johns. The typical form of the species, which is a glabrous plant, is not found in Michigan. Our plants belong to this pubescent variety, which, as pointed out by Butters and St. Johns, is *L. decaphyllus* of Pursh, but not of Britton. Keweenaw Co., June 30, 1886, No. 402a; Orion, May 30, 1895, No. 402c; Monroe, June 9, 1921, No. 5831½.

PONGELION GLANDULOSA (Desf.) Pierre. The Tree of Heaven or Ailanthus was found well established at Washington. It establishes itself in vacant lots and other places in Detroit, and persists until destroyed by improvements. Washington, Farwell and Gladewitz, June 21, 1921, No. 5882; Detroit, Oct, 19, 1893, No. 1448.

ACALYPHA VIRGINICA L. forma **purpurea**, n. f. Stems, inflorescences and under side of the leaves purple. This is a color form of a weed very common in these localities and is worth putting on record. On grassy knolls in thin copses at Washington, Farwell and Gladewitz, Sept. 14, 1921, No. 6002.

Euphorbia glyptosperma Engelm. This species was reported from Farmington in the *19th Annual Report*, p. 259; it has not been detected away from railway roadbeds and seems to be becoming more frequent in such localities. Rochester, Aug. 4, 1912, No. 2967; Washington, Gladewitz and Farwell, Sept. 14, 1921, No. 5991. Often forming dense mats three dm. or more in diameter.

Euphorbia hypericifolia L. (*E. nutans* Lag. and *E. Preslii*, Guss.) Usually along railroad banks, but frequently found in wild situations; in the former places it is mostly low and the branches are widely spreading; in the latter, the stems are erect, often a metre in height with ascending branches, especially if growing on moist banks of streams amongst tall herbage. La Salle, Gladewitz and Farwell, Sept. 21, 1921, No. 6025; Detroit, Aug. 13, 1894, No. 1478; Rochester, Aug. 4, 1912, No. 2966; Grosse Isle, Aug. 20, 1916, No. 4382.

Euphorbia dentata Mx. This is not listed in the *Michigan Flora*. It has been seen only in Wayne Co. It seems to have disappeared from Stony Island, but is on the mainland in and around Trenton. Stony Island, Sept. 18, 1909, No. 2137; Trenton, Gladewitz and Farwell, July 21, 1921, No. 5948.

Euphorbia obtusata Pursh. For the first time in many years I found this species, which is rare in Michigan. It was flourishing in great abundance in an old disused war-time garden near La Salle. Detroit, June 4, 1895, No. 971; La Salle, May 19, 1921, No. 5760.

Vitis Labrusca L. In the *6th Annual Report*, p. 208, *V. rotundifolia* Michx. was reported as found at Rochester, but the plant so determined is *V. Labrusa* L. July 4, 1896, No. 1533.

Vitis cinerea Engelm. A vine with the young angled branchlets and the young leaves densely white-woolly all over; the leaves become glabrous above but the branchlets and petioles remain more or less sparsely woolly. It was found along a fence row near La Salle. The plant had not yet flowered, but it probably belongs here. Gladewitz and Farwell, June 21, 1921, No. 5876.

HYPERICUM MACULATUM Walt. The typical form of this species with sessile, cordate-ovate, clasping, obtuse leaves is rarely met with in Michigan. It is more apt than not to be found on dry hillsides or similar situations. Detroit, July, 1904, No. 1899. The common form in moist woods, on river banks, etc., has ovate-oblong to narrowly oblong, obtuse or retuse leaves which are rounded or cuneate at base, but never cordate nor clasping, and are either sessile or distinctly petiolate. This is var. *corymbosum* (Muhl.) Farwell (*H. corymbosum* Muhl.; *H. subpetiolatum* Bickn.). The narrowly oblong, cuneate leaves strikingly resemble those of *H. prolificum.* La Salle, Sept. 21, 1921, No. 6016; Keweenaw Co., Aug. 15, 1884, No. 172; Detroit, July 16, 1892, No. 172a; Parkedale, Aug. 4, 1912, No. 2997 and Aug. 26, 1915, No. 4015½.

CALCEOLARIA CONCOLOR (Forst.) O. K. The green violet is scarce and confined to the southern half of the Lower Peninsula. It is found in moist woods. In the *Michigan Flora* it is made a synonym of the nodding violet, *C. verticillata* (Ort.) O. K., but on what grounds I can not say, since they are amply distinct. The latter is native from Colorado to Mexico and probably has never been found in Michigan. Our plant is sometimes considered to be of a different genus, *Cubelium concolor* (Forst.) Raf. Near Farmington, Sept. 28, 1921, No. 6060; Geddes, Aug. 21, 1909, No. 2123½.

The violets are very common and many of them quite variable. As I have found them in Michigan, they are as follows:

VIOLA PEDATA L. In the dry soil of hillsides. Orion, May 30, 1895, No. 986; Rochester, May 4, 1913, No. 3321; Detroit, May 19, 1902, No. 986a.

V. PALMATA L. In dry soil in woods or along fence rows in fields and along roadsides. Rochester, June 1, 1909, No. 2076; Grosse Isle, May 18, 1913, No. 3360; Utica, June 6, 1916, No. 4168.

V. TRILOBA Schwein. In dry fields under oaks. Only a few of the leaves at flowering time are lobed; the greater part are entire; the middle lobe is very broad, the leaf a perfect duplicate of the illustration in Britton and Brown. Some of the peduncles, however, are longer than the leaves; otherwise it agrees fully with the description of this species. Near Ypsilanti, Walpole, Billington and Farwell, May 30, 1920, No. 5464.

V. DOMESTICA Bicknell var. **communis** (Pollard) n. comb. (*V. communis* Pollard, *Bot. Gaz.*, 36, 336. 1898; *V. pratincola* Greene, *Pittonia* IV. 64. 1899; *V. papilionacea,* Pursh, *ex* Greene, *l. c.* [1900] 140 non Pursh, 1814; *V. familiaris* Greene, *Cybele Columb.*, [1914] 14).

This is our common smooth woodland violet and the name *V. communis* is most appropriate, but Pollard's name, given to the specialized form of restricted range, is the oldest name for the species. Parkedale, May 19, 1912, No. 2549; and May 4, 1913, Nos. 3312, 3324, 3325, 3326; Rochester, May 17, 1914, No. 3622; Stony Creek, May 4, 1913, No. 3334; Detroit, May 3, 1913, Nos. 3305, 3306 and 3308; Rockwood, May 25, 1916, No. 4145; Detroit, July 16, 1892, No. 56a; Keweenaw Co., Aug. 20, 1883, No. 56½; La Salle, May 19, 1921, No. 5767. A form is not uncommon with flowers white or mostly white and this may be known as forma **alba** (T. & G.), n. comb. (*Viola cucullata,* var. *alba* T. & G., *Fl. N. A.*, I. 137. 1838.) Parkedale, May 4, 1913, No. 3311 and May 17, 1914, No. 3628; Rockwood, May 21, 1916, No. 4144.

V. MISSOURIENSIS Greene. In alluvial soil. The flowers were essentially white; otherwise it agrees fully with the description of this species. Erie, May 10, 1921, No. 5725.

V. SORORIA Willd. Our common pubescent woodland violet. Detroit, Sept. 31, 1897, No. 1602; Rochester, June 1, 1909, No. 1602a; Parkedale, May 19, 1912, No. 2551; Grosse Isle, May 18, 1913, No. 3358. The next three are generally considered to be synonymous with this.

V. CUSPIDATA Greene. Rochester, May 10, 1914; Nos. 3609, 3610, and June 7, 1914, No. 3656.

V. NODOSA Greene. Parkedale, May 17, 1914, No. 3631; Grosse Isle, May 18, 1913, No. 3359.

V. DICKSONII Greene. Rochester, May 10, 1914, No. 3613.

V. POPULIFOLIA Greene, also considered to be a synonym of

V. sororia, was collected in southern Michigan by Greene, but I have not seen it. *V. villosa* of the *Michigan Flora* is *V. renifolia.*

V. OBLIQUA Hill. (*V. affinis* Le Conte and *V. venustula* Greene). In moist woods, borders of streams, etc. Keweenaw Co., Aug. 20, 1883, No. 56⅓ and Aug. 25, 1898, No. 1619a; Rochester, May 10, 1914, No. 3612.

V. CRENULATA Greene. This, also, is generally considered to be a synonym of the preceding. Rochester, May 10, 1914, No. 3618; Walled Lake, June 5, 1917, No. 4472.

V. SEPTENTRIONALIS Greene. Keweenaw Co., May, 1908, No. 2048.

V. CUCULLATA Ait. (*V. papilionacea* Pursh, *Fl. Am.*, Sept. I, 173. 1814). This is our common bog meadow violet. Keweenaw Co., Aug. 20, 1883, No. 56, Aug. 25, 1898, No. 1619 and Sept. 20, 1914, Nos. 3908⅓ and 3908⅓; Rochester May 10, 1914, No. 3614; Parkedale, May 19, 1912, No. 2565, June 9, 1912, No. 2637, May 30, 1914, No. 3654, May 10, 1914, Nos. 3619 and 3620; May 17, 1914, No. 3630. In *Cybele Columbiana*, I. 26–27, 1914, the late Dr. Greene mentions having found a living, flourishing violet with yellow beard on the petals. He stated that when the time came he would give it a new name rather than apply to it the specific name *papilionacea* of Pursh. On the occasion when he found it he was my guest at Detroit and I was taking him for a jaunt through the violet fields around Rochester and Parkedale. The plants with the yellow bearded petals were the same as my numbers 3619 and 3620. These were plants that I particularly wished to show him and they were the main reason for his being there. He found a convenient log on which to sit and spent much time with a lens and rule making and entering memoranda in his note-book and giving the plants the provisional MS. name of *V. caricetorum* because they were found in a Carex bog. The bog was an extensive one and of marl formation. I wish to make the statement, however, that the "beard" was not different in color from that of other violets. The compound microscope showed that the apparently yellow color of the beard was due to empty pollen sacs that had fallen and had been caught and entangled in the beard. I presume this condition may occur in any species, but I have not happened to detect it in any other.

V. CRASSULA Greene. Parkedale, May 30, 1914, No. 3653; Rochester, May 17, 1914, No. 3623; May 26, 1914, No. 3644; May 30, 1914, No. 3647.

V. PERAMOENA Greene. May 18, 1913, No. 3370; Rochester, May 10, 1914, No. 3616.

V. NEPHROPHYLLA Greene. Algonac, May 2, 1915, No. 3946.

V. VAGULA Greene. Parkedale, May 19, 1912, No. 2550 and May 25, 1913, N. 3413. It was very plentiful in a marl bog. As in the case of *V. caricetorum*, mentioned above, Dr. Greene found a convenient seat and spent much time with lens, rule, paper and pencil. When he had finished his thorough examination, he declared it to be a new species which in due time he would publish as *V. Farwellii*. He acknowledged that it was closely related to his *V. vagula*, but quite different.

V. SAGITTATA Ait. A glabrous violet with arrow-shaped leaves found in marshes, on the borders of ponds, streams, etc., often in moist sand. Rare. Keweenaw Co., June 26, 1884, No. 106; Ypsilanti, May 16, 1891, No. 106b. A larger pubescent form is var. *subsagittata* (Greene) Farwell. In sandy fields near Bloomfield Center, Sept. 16, 1917, No. 4605; Detroit, May 1, 1919, No. 5223a; Portage Lake, June 12, 1921, No. 5851.

V. FIMBRIATULA J. E. Sm. A similar species but with more ovate, less toothed leaves, more frequently preferring woods and thickets. Detroit, May 1, 1896, No. 862; Algonac, May 27, 1912, No. 2584.

V. EMARGINATA (Nutt.) Le Conte. Our plants are found in sandy soils, sometimes with no other vegetation, sometimes with a sparse growth of sedge or grass. These plants are never entirely free of pubescence and never show the broad deltoid leaf characteristic of this species, but Dr. Greene said that they were nevertheless unmistakably this species. Royal Oak, June 17, 1916, No. 4211; Detroit, May 14, 1898, No. 1603; Algonac, May 24, 1914, Nos. 3642 and 3643.

V. SELKIRKII Pursh. Rich cool woods, Keweenaw Co., June 26, 1884, No. 104; Lake Linden, Sept. 27, 1914, No. 3908¼.

V. RENIFOLIA A. Gr. In rich cool woods. Keweenaw Co., May 15, 1884, No. 81 and June 6, 1884, No. 106a; Lake Linden, Sept. 27, 1914, No. 3908½. A form with upper leaf surface glabrous is var. *Brainerdi* (Greene) Fernald. The variety is the form growing in dryer and warmer situations along the borders of woods or out in the open. Lake Linden, June 25, 1915, No. 3991.

V. INCOGNITA Brainerd. In moist, grassy copses. Lake Linden, Sept. 27, 1914, No. 3906; Keweenaw Co., May 15, 1884, No. 82.

V. BLANDA Willd. Detroit, May 9, 1895, No. 82a; Stony Creek, May 10, 1914, No. 3621; Algonac, May 2, 1915, No. 3948; Utica, June 6, 1916, No. 4175; Lake Linden, Sept. 30, 1914, No. 3908a.

V. PALLENS (Banks) Brainerd. Keweenaw Co., July 6, 1883, No. 26; Detroit, June 2, 1896, No. 26a; Stony Creek, May 14, 1913, No. 3336; Walled Lake, June 5, 1917, No. 4476.

V. LANCEOLATA L. Only once collected on Belle Isle, June 2, 1896, No. 869.

V. PUBESCENS Ait. Our common yellow violet; the typical form has pubescent stems and leaves, but glabrous capsules. Keweenaw Co., July 10, 1883, No. 28 and Oct. 2, 1914, No. 3908b; Detroit, May 6, 1893, No. 28a and May 3, 1913, No. 3307; Stony Creek, June 28, 1914, No. 3706. Another form with woolly capsules should be known as *Viola pubescens* Ait. forma **eriocarpa** (Nutt.) n. comb. *Viola pubescens* var. *eriocarpa* Nutt., *Gen.* I. 150. 1818. With the type but much less common. Keweenaw Co., Aug. 20, 1883, No. 57; Detroit, May 19, 1893, No. 57a; Parkedale, May 25, 1913, No. 3410; Stony Creek, June 8, 1913, No. 3442 and June 28, 1914, No. 3707; Pittsfield, May 15, 1921, No. 5741. A third form with glabrous pods differs in being glabrous or slightly scabrous on the upper parts rather than pubescent.

VIOLA PUBESCENS Ait. var. SCABRIUSCULA T. & G. forma **leiocarpa** (Fernald & Wiegand) n. comb. (*Viola eriocarpa* Schw. var. *leiocarpa* Fernald & Wiegand in *Rhodora,* 23: 275. 1922.) Rochester, June 1, 1909, No. 2077; Parkedale, May 19, 1912, No.

2554 and May 25, 1913, No. 3404; Redford, May 24, 1918, No. 4860. Keweenaw Co., Oct. 19, 1914, No. 3908d; Lake Linden, Oct. 18, 1914, No. 3908c. The capsules of this variety are said to be either glabrous or woolly. The Michigan plants that I have seen have had glabrous capsules. This form also differs from the other two in being branched from the base so that there are several stems in a clump; the others have the stems unbranched or single. All these forms are, however, manifestly of the same species.

V. Canadensis L. Essentially glabrous. Detroit, June 2, 1896, No. 868; Rochester, June 1, 1909, No. 868a; Washington, May 25, 1916, No. 4154½. A form is hirtellous throughout and this is var. *pubens* Farwell. Washington, May 25, 1916, No. 4155.

V. striata Ait. Keweenaw Co., May 15, 1884, No. 80; Utica, June 6, 1916, No. 4164; Ypsilanti, May 30, 1920, No. 5456; Redford, May 24, 1918, No. 4862.

V. conspersa Reichb. Keweenaw Co., July 10, 1883, No. 27; Detroit, April 30, 1895, No. 27a; Parkedale, Oct. 27, 1912, No. 3260; Utica, June 6, 1916, No. 4165; Algonac, May 2, 1915, No. 3950; Royal Oak, June 17, 1916, No. 4212; Lake Linden, Sept. 27, 1914, No. 3905. A form with white flowers is var. *Masoni* Farwell, June 6, 1916, No. 4163; Redford, May 24, 1918, No. 4865.

V. subvestita Greene. (*V. arenaria* Amer. authors, not DC.) Dry soils in Keweenaw Co., July 23, 1883, No. 39.

V. rostrata Pursh. Ypsilanti, May 21, 1892, No. 1233; Detroit, May 9, 1895, No. 1233a; Birmingham, May 18, 1902, No. 1233b; Rochester, May 12, 1909, No. 1233c; Parkedale, May 19, 1912, No. 2548; Rockwood, May 25, 1916, No. 4147. In the typical form of the species the spur is as long as the petals, about 12 mm.; var. *elongata* Farwell has the spur much longer, 15–18 mm., also narrower sepals. Utica, June 6, 1916, No. 4166. A white-flowered form is var. *Phelpsiae* (Fernald) Farwell. Rockwood, May 25, 1916, No. 4146.

V. tricolor L. The common pansy has escaped in places. Keweenaw Co., June 26, 1884, No. 105; Detroit, June 12, 1894,

No. 105a. A more slender form with small flowers is var. *arvensis* (Murr.) DC. Dry fields. Utica, June 6, 1916, No. 4162; Portage Lake, Walpole, June 11, 1921 and Walpole, Billington, Gladewitz and Farwell, June 12, 1921, No. 5852. A form of this with leaves of a rounder shape and with the terminal section of the stipules more entire, may be known as forma **tenella** (Muhl.) n. comb. (*V. tenella* Muhl., *Cat.*, 26. 1813; *V. Rafinesquii* Greene *Pittonia* IV. 9. 1899.) Dry fields at Detroit, Aug. 22, 1893, No. 945.

OENOTHERA PRATENSIS (Small) Robinson. This pretty evening primrose is quite frequent at Washington. The whole plant is very hirsute and many of the capsules are sterile. Gladewitz and Farwell, June 21, 1921, No. 5887. Our local manuals exclude Michigan from the range of this species, but southern Michigan, at least, is within its region of distribution.

THASPIUM BARBINODE (Mx.) Nutt. var. ANGUSTIFOLIUM Coult. & Rose. Along the railroad tracks and banks near Monroe. Gladewitz and Farwell, June 9, 1921, No. 5832.

DAUCUS CAROTA L. f. **epurpurata**, n. f. The purple central floret of the umbell and umbellets is absent; flowers white. Dundee, July 15, 1919, No. 5305.

CORNUS ASPERIFOLIA Michx. Our local manuals call this species a shrub, but it is often a tree with a trunk 8 inches in diameter near the ground and 20 feet or more in height. Banks of streams in alluvial grounds near La Salle, Gladewitz and Farwell, Sept. 21, 1921, No. 5764. The shrubby form was collected at Detroit, Sept. 16, 1892, No. 1336.

VACCINIUM PENNSYLVANICUM Lam. var. MYRTILLOIDES (Michx.) Fernald. The form of the common low blueberry with pubescent leaves and branchlets, Detroit, April 8, 1921, No. 5718.

V. VACILLANS Kalm forma **rosea**, n. f. Corollas red, — as red as in *Gaylussacia baccata*. A color variation that seems to be permanent enough to name and put on record. Detroit, April 8, 1921, No. 5719.

OXYCOCCUS OXYCOCCOS (L.) Mac M. var. **intermedius** (A. Gray), n. comb. (*Vaccinium oxycoccos* L. var. *intermedium* A. Gray, *Syn. Fl.*, II. pt. 1, 396. 1886.) A form with more numerous and larger flowers and larger leaves which, as the name indicates, is more or less intermediate between this species and *O. macrocarpus.* Bloomfield Center, June 15, 1916, No. 4192.

Samolus floribundus HBK. The water pimpernel is not a common plant in Michigan. It is recorded only for a few places. Mr. Chandler discovered it on the banks of a stream in Wayne Co., not far west of Trenton, some five or six years ago. At that time there were only a dozen or so of the plants. Now it has spread far up and down the stream and is found there in great abundance. Gladewitz & Farwell, July 26, 1921, No. 5930.

Asclepias tuberosa L. Pleurisy root. Linnaeus described his species as with lanceolate leaves and orange flowers. The common form in Michigan has alternate, ovate, lance-ovate or lance-oblong leaves, 5–10 cm. in length, widest at the base where it is 1–3 cm. in width, and usually subcordate, acutish at the apex, stems erect. This is var. *cordata* of Eaton & Wright. Ypsilanti, June 23, 1891, No. 1157; Detroit, July 6, 1900, No. 1157a; Rochester, July 30, 1912, No. 2760; Oxford, Aug. 22, 1920, No. 5618; Utica, Oct. 5, 1921, No. 6088.

Var. *decumbens* Pursh. (Not *A. decumbens* (L.) Willd, etc.) Although Pursh cited *A. decumbens* Willd. as a synonym, his description manifestly shows that the plant described could not be the species of Willd. Evidently it was the *A. tuberosa* var. β of Michx. It is decumbent and the leaves are much narrower than in the preceding variety, approaching the linear; occasionally narrowly oblanceolate, subcordate, rounded or tapering at the base. Oxford, Aug. 22, 1920, No. 5617a. I have not seen the typical, erect form with lanceolate leaves except in the form with yellow flowers which may be known as forma **flavescens**, n. f. Rochester, June 21, 1921, No. 5875.

Convolvulus incanus Vahl var. **glabratus**, n. var. Canescence so sparse that the plant appears to be green. Detroit, July 21, 1921, No. 5950. The typical form of the species is densely canescent; the plant appears to be white. Undoubtedly this is an introduction from farther west or southwest and is new to Michigan.

VERBENA HASTATA L. Moistly simple about 1.3 m. or less in height, with a not very large panicle, the rather stiff spikes of which are thick and densely flowered; leaves usually hastate. This is the common form widely distributed. Detroit, July 21, 1892, No. 1265; Parkedale, July 28, 1912, No. 2905; Franklin Sept. 23, 1918, No. 5158. Var. *paniculata* (Lam.) Farwell is much taller, often over 2 m. high, very much branched with longer and more slender, flexuous spikes, and usually with but few of the lowermost leaves hastate. Usually along river banks; infrequent. Detroit, July 15, 1899, No. 1265a; Rochester, July 4, 1901, No. 1265b; Trenton, July 26, 1921, No. 5940. A form of this variety with clear pink flowers may be known as forma **rosea**, n. f. Trenton, July 26, 1921, No. 5939. A probable hybrid between *V. hastata* and *V. urticifolia* is *V. hastata* var. *oblongifolia* Nutt. The leaves are narrow, some of them hastate, the panicles are of the type of *V. urticifolia*, but the flowers are bluish at the base and pink above. Detroit, Aug. 21, 1909, No. 2126; Franklin, Sept. 23, 1918, No. 5139; South Rockwood, July 12, 1921, No. 5927.

[LEONURUS MARRUBIASTRUM Linn.[2] This species of motherwort is recorded in our local manuals as naturalized from Europe on waste grounds in New Jersey, Pennsylvania and Delaware. I found it in Chicago where it nearly covered an entire vacant, unimproved lot. July 16, 1920, No. 5547. Since this species is traveling westward, it may be found in or near the larger cities of Michigan.]

Koellia Virginica (L.) n. comb. *Satureja Virginica* L., *Mant.*, II. 40. 1771; *Brachystemum linifolium* Willd., *Enum.* 623. 1809; *Pycnanthemum linifolium* Pursh, *Fl. Am. Sept.*, II. 409. 1814; *K. flexuosa* (Walt.) MacM., *Metasperm. Minn.*, 452. 1892. Also the following as to name-bringing synonym and description in small part: *Thymus Virginicus* (L.) Murray, *Syst. Veg.* (ed. 14), 542. 1784; Gmelin, *Syst. Nat.* (ed. 13), reform., II, pt. 2, 913. 1796; Willd., *Sp. Pl.* III. 145. 1800; and *Brachystemum Virginicum* (L.) Mx., *Fl. Bor. Am.*, II. 6. 1803. Probably also, at least in part, and as to name-bringing synonym *Nepeta Virginica*

[2] Not a member of the Michigan flora.

L., *Sp. Pl.*, II. 571. 1753 and I. 799. 1762; *Syst. Nat.* (ed. 13), II. pt. 1, 390. 1770; *Mill. Dict.*, No. 9. 1768; Murray, *l. c.* 530; Gmelin, *l. c.*, 900; Willd., *l. c.*, 56.

THYMUS VIRGINICUS L., *Mant.*, II. 409, is listed in the *Index Kewensis*, IV. 1076 (1895) and referred to *Pycnanthemum lanceolatum* and *Satureja Thymus Virginicus* L., *Mant.* II. 409 (1771), is listed in both editions of Britton and Brown as a synonym of *Koellia flexuosa* (Walt.) MacM. Linnaeus is not correctly transcribed in either publication. "Thymus" is not the generic name used by Linnaeus as that genus is listed on page 413 of the *Mantissa;* nor is it a part of the specific name as it is not on the margin of the page where both generic and specific names occur, but is a part of the descriptive paragraph; at the place cited Linnaeus published *Satureja Virginica* and probably was only transferring his earlier *Nepeta Virginica* to *Satureja.* The *Index Kewensis*, Part III. 307 (1895), lists *Nepeta Virginica* Willd., *Sp. Pl.*, III. 56, and refers it to *Pycnanthemum lanceolatum* and retains *Nepeta Virginica* L., *Sp. Pl.*, 571, as a valid species. In view of the fact that Willdenow merely copied the description, synonym and note from Linnaeus, it may puzzle most people to know on what grounds it is considered to be something else. Also no valid species of catnip has been recognized as *Nepeta Virginica* by American authors in more than a century, if it ever was. So far as I am able to discover, Murray is the real author of the combination *Thymus Virginicus*, but it is evident that he included under this name both the *Satureja Virginiana* and the *S. Virginica* of Linnaeus, as his description combined the descriptions of both. He was followed by Gmelin, Willdenow and Michaux. The two were separated again by Willdenow in the *Enumeratio*. Michaux quoted *Nepeta Virginica* L. as a synonym of his *Pycnanthemum aristatum* and was followed by Eaton & Weight, since which time it has disappeared from American botanical literature. Article 57 of the Vienna Rules provides that "When the difference between two names . . . lies in the termination, these names are to be regarded as distinct even though differing by one letter only;" therefore the specific name *Virginica* must be considered as distinct from

Virginiana and must replace *flexuosa* (Walt.) MacM. This species is essentially glabrous; it has linear leaves and subulate calyx lobes; it is rare in Michigan. Detroit, Aug. 13, 1904, No. 1848 and Aug. 7, 1910, No. 1848b; Royal Oak, Aug. 6, 1910, No. 1848a.

Solanum Dulcamara L. The true bittersweet which is quite common both in wild locations and near habitations. Berries are said to be poisonous. Leaves rather small, 4 cm. long, and broadly ovate with cordate base, 3 or 4 cm. wide, or narrowly ovate, 2–2.5 cm. wide, with a rounded or narrowed base, all entire except the very uppermost which are hastately lobed. Whole plant glabrous. Flowers purple. Ypsilanti, June 13, 1891, No. 1148; Detroit, June 1, 1894, No. 1148b.

Var. *pubescens* R. & S. (*S. littorale* Raab; *S. Dulcamara* var. *tomentosum* Koch; and vars. *laciniatum* and *hirsutum* Dun.) The whole plant is hirsutely pubescent, the leaves larger and more generally lobed, some of them 5- to 7- lobed or even foliolate; flowers usually blue. Detroit, July 25, 1892, No. 1148a; Parkedale, June 23, 1912, No. 2735; Orion, July 7, 1918, No. 5056; Trenton, July 26, 1921, No. 5943. A form with very sparse pubescence and white flowers may be known as forma **albiflora,** n. f. Trenton, July 26, 1921, No. 5944.

Var. **canescens,** n. var. Canescent, flowers white. Rochester, Aug. 15, 1909, No. 2105. I have not seen this variety except in low, wet grounds or along banks of streams and always in shady woods. Rare.

Linaria minor (L.) Desf. A small, blue-flowered species first reported from Michigan by me. It is now very common along railroad tracks all through southeastern Michigan. Usually simple and about 1 dm. in height, but occasionally 3 or 4 dm. high and copiously branched. Detroit, June 24, 1894, No. 1463½; Rockwood, June 16, 1918, No. 4949; Stony Creek, June 9, 1921, No. 5818.

Pentstemon Digitalis (Sweet) Nutt. Adrian is the only locality given in the *Michigan Flora*. In addition to Detroit and Algonac, previously reported, this has been found near Royal Oak, Aug. 6, 1910, No. 1916a; and by Gladewitz and

Farwell at Stony Creek, June 9, 1921, No. 5822; and at Washington, June 21, 1921, No. 5894. It is plentiful wherever seen, and is now frequent throughout the southeastern section of the State.

P. Pentstemon (L.) Britt. Grand Rapids is the only station given for this in the *Michigan Flora*. It was collected at Detroit by me, Aug. 3, 1907, No. 2036.

Mimulus alatus Sol. Rare in the southeastern section of Michigan. Zoölogical Park near Royal Oak, Aug. 27, 1916, No. 4412.

Veronica anagallis-aquatica L. var. latifolia (Britt.) (as V. *Anagallis* var. *latifolia*). Eagle Harbor, in slow, muddy streams, July 8, 1915, No. 4005. The plant is succulent and has very broad, fleshy leaves; sepals acute or occasionally somewhat obtuse. In the type the sepals are acuminate. As Dr. Britton originally stated, there are no diagnostic characters in the pedicels and capsules. Another station has been found for var. *glandulosa* Farwell. It is quite plentiful in the stream near Trenton where Samolus grows. Gladewitz and Farwell, July 26, 1921, No. 5931. In this variety the upper part of the stems and inflorescence are minutely glandular and the calyx lobes are oval and obtuse.

Plantago spinulosa Dcne. A smaller species than *P. aristata*, but similar to it, yet readily distinguished by its lighter green color, shorter bracts, filiform leaves, absence of the dense white woolly pubescence about the apex of the scape and inflorescence, and by its earlier flowering period, which has passed before that of *P. aristata* has begun. In this the pubescence is mostly appressed while in the other it is mostly spreading, and, as stated above, does not have the dense white woolly appearance seen in the latter, but on the contrary presents the appearance of a glabrate or glabrous plant. Associated with *P. aristata* at Franklin, Billington, Gladewitz and Farwell, June 24, 1920, No. 5504.

P. aristata Michx. This plantain is becoming quite common in dry, sandy or sterile soil. Small specimens of it are readily distinguished from the preceding by their darker green color and by their contrasting white woolly pubescence, longer bracts and later flowering period. Rochester, June 30, 1909; Zoölogical Park near Royal Oak, June 17, 1916, No. 4214; Franklin, Billington, Gladewitz and Farwell, June 24, 1920, No. 5503. At this time this species was only in bud while the associated *P. spinulosa* was approaching the close of its flowering season. Portage Lake, Walpole, Billington, Gladewitz and Farwell, June 12, 1921, No. 5853 (very young; early bud stage).

P. Virginica L. Dry pastures near Portage Lake, Walpole, June 11, 1921; Walpole, Billington, Gladewitz and Farwell, June 12, 1921, No. 5856.

Valeriana officinalis L. Cultivated and inclined to escape. Trenton, May 26, 1921, No. 5794.

Valerianella radiata (L.) Dufr. In swampy grounds near La Salle where it covers acres of land. Gladewitz and Farwell, May 19, 1921, No. 5768. A variety with smooth fruit, *leiocarpa* (T. & G.) Krok., is as common as the typical form. June 2, 1921, No. 5813.

Sambucus Canadensis L. var. laciniata A. Gray. A form in which the leaflets are ternately divided. Near Farmington, Sept. 28, 1921, No. 6049.

Viburnum dentatum L. This species is listed in the *Michigan Flora*, but the only station given is Sarnia, Ontario; C. K. Dodge. I collected it at Rochester, Oct. 25, 1914, No. 3920; and at Pittsfield, in bud only, May 15, 1921, No. 5740.

Campanula Carpathica Jacq. An escape from cultivation at Eagle River; Aug. 16, 1921, No. 2971.

Eupatorium perfoliatum L. var. truncatum (Muhl.) A. Gray. In this variety the leaves are separated and truncate or rounded at base. On some plants all the leaves are separated, but more often only the upper ones. With the species, but much less common. Slocum's Island, Aug. 31, 1921, No. 5979a; Washington, Gladewitz and Farwell, Sept. 14, 1921, No. 6000.

Aster macrophyllus L. One of our common and variable asters, the type of which is common enough throughout the State. Keweenaw Co., Sept. 10, 1886, No. 478; Orion, Aug. 19, 1895, No. 478a; Lakeville, Sept. 2, 1901, No. 478b; Goodison,

Oct. 7, 1917, No. 4684. Some of its subdivisions and segregates are mentioned herewith: var. *velutinus* Burgess, Lakeville, Sept. 2, 1901, No. 1749; Copper Harbor, Oct. 7, 1910, No. 1749a; Washington, Sept. 14, 1921, No. 5995; var. *sejunctus* Burgess, Copper Harbor, Oct. 7, 1910, No. 2108a; Rochester, Aug. 15, 1909, No. 2108; Detroit, Sept. 14, 1912, No. 3171; var. *apricensis* Burgess, Copper Harbor, Oct. 7, 1910, No. 2197; var. *excelsior* Burgess, Keweenaw Co., Sept. 15, 1889, No. 731.

A. ROSCIDUS Burgess. Copper Harbor, Oct. 7, 1910, No. 2198; Oxford, Oct. 11, 1917, No. 4734.

A. IANTHINUS Burgess. Royal Oak, Aug. 6, 1910, No. 2175; Oxford, Oct. 11, 1917, No. 4724; Slocum's Island, Aug. 31, 1921, No. 5984.

A. MULTIFORMIS Burgess. Rochester, Oct. 5, 1913, No. 3524.

A. NOBILIS Burgess. Rochester, Aug. 15, 1909, No. 2113; Marl Lake, Sept. 18, 1920, No. 5712.

A. SHORTII Hook. Washington, Gladewitz and Farwell, Sept. 14, 1921, Nos. 5997 and 5996. Most of the leaves in the latter number were very scabrous above, where the pubescence was almost as copious as in *A. azureus*, but much shorter and and more harsh. This species was also collected by C. Billington at Tecumseh, September 5, 1920.

A. LAWRIEANUS Porter. This was found quite plentifully in 1921. Lakeville, Sept. 2, 1901, No. 1745; Oxford, Oct. 11, 1917, No. 4738; Washington, Sept. 14, 1921, No. 6008, Farmington, Sept. 28, 1921, No. 6048.

A. SAGITTIFOLIUS Willd. The typical form of the species has an open paniculate inflorescence, with blue rays; it is widely distributed, but not common. Ypsilanti, Aug. 5, 1891, No. 1192½; Detroit, July 31, 1909, No. 1192¼; Parkedale, Sept. 2, 1912, No. 3110; Marl Lake, Oct. 11, 1917, No. 4711½. Var. *dissitiflorus* Burgess; Ypsilanti, Aug. 5, 1891, No. 1192⅓; Rochester, Oct. 5, 1913, No. 3525.

Var. *urophyllus* (Lindley) Burgess. This is the commonest form of the species in southeastern Michigan. It differs from the others in having more contracted panicles, pedicels short and concealed by the numerous heads which form a thyrse;

rays white; petioles and midveins usually pilose. Detroit, Sept. 7, 1892, No. 1329; Parkedale, Aug. 23, 1914, No. 3841; Rochester, Aug. 23, 1914, No. 3835 and Sept. 7, 1914, No. 3853. If the stems have been cut down, the new branches and leaves are very apt to be entirely glabrous. Grosse Isle, H. C. Hamilton, Oct. 25, 1921.

ERIGERON RAMOSUS (Walt.) BSP. var. **integrifolius** (Bigel.), n. comb. (*Erigeron integrifolium* Bigel., *Fl. Bost.*, 302. 1824.) The typical form of the species is gray throughout from the copious appressed pubescence. The variety is green; the pubescence is very sparse. As the two forms grow here the variety is also larger and more branched. It appears to be as distinct as *E. annuus*. Portage Lake, B. A. Walpole, June 11, 1921; Walpole, Billington, Gladewitz, and Farwell, June 12, 1921, No. 5848; Washington, June 21, 1921, No. 5880.

AMBROSIA PSILOSTACHYA DC. Rapidly spreading. Keweenaw Co., August, 1902, No. 1784; Newaygo Co., where it is very common around Woodville, Aug. 4, 1921, No. 5956; Houghton Co., Aug. 10, 1921, No. 5964.

RIDAN ALTERNIFOLIUS (L.) Britton var. **oppositifolius** (Fresen.), n. comb. (*Actinomeris squarrosa* var. *oppositifolia* [Fresen.] T. & G., *Fl. N. Amer.*, II. 335. 1842.) The typical form of the species with alternate leaves is very common in the southeastern section of the State. Detroit, Sept. 27, 1906; Dearborn, July 8, 1917, No. 4517; Monroe Piers, Sept. 9, 1917, No. 4565; Redford, Oct. 14, 1917, No. 4744; La Salle, Sept. 21, 1921, No. 6028. The variety with opposite leaves was found at La Salle, Sept. 21, 1921, No. 6029.

ACHILLEA BOREALIS Bongard. Copper Harbor, Aug. 16, 1921, No. 5968. Larger and coarser than *A. Millefolium*. The involucral bracts have a very dark brown border.

MATRICARIA SUAVEOLENS (Pursh.) Buchenau. This species of chamomile is a native of the Pacific Coast, but has migrated eastward. Calumet, Aug. 17, 1921, No. 5973. A vacant lot was completely overrun with it to the exclusion of everything else.

ARTEMISIA CAUDATA Mx. Frequent in Newaygo Co., Woodville, Aug. 4, 1921, No. 5961.

A. BOREALIS Pall. var. WORMSKIOLDII Bess. A very rare plant 4 dm. in height and very silky villous throughout. Eagle Harbor, Aug. 16, 1921, No. 5969.

LACTUCA CANADENSIS L. Linnaeus described his species as follows:

canadensis 4. LACTUCA foliis lanceolato-ensiformibus dentatis inermibus.

Lactuca canadensis altissma angustifolia, flore pallide luteo. Tournef., *Inst.*, 474.

Sonchus sylvestris, folio laciniato glauco; costa non spinosa. *Raj. Suppl.* 137.

Habitat in Canada. *Kalm.*

Differt a sativa caule altiore; foliis *quadruplo longioribus, angustioribusque;* racemo *terminali longo, composito, nec corymboso.*

There is nothing in the description above that will indicate a plant with pinnatifid leaves; certainly our common wild lettuce with runcinate pinnatifid leaves can not by the wildest stretch of imagination be denominated a plant with "foliis lanceolato-ensiformibus." Linnaeus compares the leaves with those of *L. sativa*, a species with *entire* leaves. He would not have done this if the leaves were pinnatifid. Undoubtedly the species is based upon Tournefort, from whose description the specific name, *Canadensis,* is derived; Tournefort also describes a plant with entire leaves. The quotation from Ray is the only indication of a pinnatifid leaf, but as the specific name was derived from Tournefort, this must be considered as the type, and is equivalent to what is known as *C. intergrifolia* Bigel., the leaves of which fully answer the Linnaean description of *lanceolate,* 'sword-shaped.' It is also the *Sonchus pallidus* Willd., *Sp. Pl.* III. 1521, 1804, and *L. sagittifolia* Ell., *Bot. S. C. and Ga.*, II. 253, 1822. The leaves are ovate-lanceolate or lanceolate and entire, or sometimes with a rounded, lobe-like enlargement on each side near the base, giving to the leaf a very good outline of a sword, whence the descriptive phrase, *lanceolato-ensiformibus;* the margins are entire, denticulate, or toothed; the lowermost are sometimes shallowly lobed. Detroit, July 21, 1892, No. 310b and July 21, 1893, No. 1400; Rochester,

July 23, 1909, No. 2094; Star, Aug. 31, 1909, No. 2094a; Lathrop, Oct. 21, 1917, No. 4784.

The Lathrop specimens had purplish stems and leaves, which may be a change of color due to the lateness of the season. The lowermost leaves are occasionally obovate or oblanceolate, showing a close relationship to var. *obovata* Wiegand, which has the leaves oblanceolate or obovate and acuminate, or rarely oblong-spatulate with a rounded apex. Trenton, July 26, 1921, No. 5946. Typically all the leaves are entire, but our plant has the lowermost leaves runcinate pinnatifid with ovate or sometimes obovate toothed lobes showing a transitional stage toward var. *latifolia* O. K., which has all the leaves pinnatifid. The lobes are ovate and usually toothed. Detroit, Aug. 3, 1899, No. 310c; Bloomfield, Aug. 22, 1916, No. 4393½; Monroe, July 7, 1921, No. 5917; S. Rockwood, July 12, 1921, No. 5923. The series listed above have broad leaves or lobes; there is another parallel series with very narrow leaves and lobes, the oldest name for which is *L. longifolia* Mx., *Fl. Bor. Amer.*, II. 85. 1803. It may be known as var. **longifolia** (Mx.), n. comb. (*L. elongata* var. *longifolia* T. & G., *Fl. N. Amer.*, II. 496. 1843.) The leaves or their lobes, the terminal one much elongated, are linear or linear-lanceolate. The leaves are often 3 dm. long and the lobes 6 cm. The lowermost are pinnatifid with 3 or 4 pairs of distant lobes passing into the entire leaves which cover the upper two-thirds of the stem. Slocum's Island, Aug. 31, 1921, No. 5981; Trenton, July 26, 1921, No. 5945; S. Rockwood, July 12, 1921, No. 5922; Monroe, July 7, 1921, No. 5915; Rochester, July 20, 1912, No. 2881; Keweenaw Co., July 26, 1886, No. 310a. A color variation of this with pale blue or white flowers may be known as forma **albocaerulea** (Farwell), n. comb. (*L. Canadensis* var. *albocaerulea* Farwell, *Mich. Acad. Sci.*, 19, 250. 1917.) Another form with all the leaves up to the inflorescence pinnatifid may be known as var. **laciniata**, n. var. Keweenaw Co., Aug. 12, 1885, No. 310; Rochester, July 14, 1912, No. 2840; Monroe, July 7, 1921, No. 5916; S. Rockwood, July 12, 1921, No. 5921. Another form, intermediate between the var. *longifolia* and var. *laci-*

niata, as to proportion of entire leaves, has smaller leaves with more closely placed lateral lobes and with the terminal lobe much reduced in length proportionately is *L. elongata* Muhl. in Willd., *Sp. Pl.* III. 1525. 1804. It, may be known as var. **elongata** (Muhl.), n. comb. Bloomfield, Aug. 22, 1916, No. 4393. The entire leaves are few and just below the inflorescence. Wiegand described and named several forms with tapering, non-sagittate leaves, but none of these have been detected here.

Department of Botany
Parke, Davis & Co.
Detroit, Michigan

EXTENSIONS OF RANGE OF PLANTS IN THE DOUGLAS LAKE REGION, CHEBOYGAN COUNTY, MICHIGAN

FRANK C. GATES

The following interesting extensions in the range of certain plants in the Douglas Lake region, Michigan, were noted in the summers of 1919 and 1920.

GRINDELIA SQUARROSA (Pursh) Dunal. In passing along a road about two years ago, a yellow-flowered plant was noted by the author and suspected to be a species of Grindelia. Opportunity to visit this place, however, was not afforded until August of 1921, when the identity was established as indicated. This plant is very common in the prairie region, but prairies do not occur in this part of Michigan. At present there are a half a dozen clumps of the plant, apparently well established and giving every evidence of permanent location. As they are on a cross-road, they probably have been brought into Pellston by trains and spread out from there.

SIBBALDIOPSIS TRIDENTATA (Soland.) Rydb. This northern plant, according to Britton and Brown, ranges south into the Upper Peninsula of Michigan. South of this, however, along the right of way of the G. R. & I. Railroad a short distance north of Pellston, there is an excellent patch of this plant from eight to ten feet in width and from forty to fifty feet in length, quite evidently flourishing and spreading.

SAGITTARIA CRISTATA Engelm. This aquatic plant, which is not given in Gray's Manual (seventh edition), but is stated in Britton and Brown (second edition) to occur in Iowa and Minnesota, was found by the author in company with Professors Ehlers and Nichols in Crooked River in the vicinity of Alanson in August of 1921, occurring as a small patch in a bend of the river in a disturbed place. It would appear that the patch was well established and it is to be expected that it will spread.

OENOTHERA OAKESIANA Robbins. Three plants with the calyx segments appendaged a third of the way below the tip and velvety pubescent were found in sandy ballast at the harbor of Cecil, July 31, 1921. These plants failed to have any thick based hairs characteristic of *Oenothera muricata* L., which is occasional in this region, but which has the appendage, if present, at the tip of the calyx lobe. This extension of range from Long Island and east is quite noteworthy, but as only three plants were found, it is probably not to be considered as established.

KANSAS STATE AGRICULTURAL COLLEGE
MANHATTAN, KANSAS

STUDIES ON THE HYDROGEN ION CONCENTRATION OF PLANT JUICES

I. Preliminary Studies on the Changes in the Hydrogen Ion Concentration of Plants during their Development

F. G. GUSTAFSON

In connection with another investigation the writer noticed that, as plants matured, the H ion concentration (i.e., the actual acidity) of the juice decreased, or, as we would say, the pH increased. This decrease in H ion concentration was sometimes quite large. In recent years it has been shown that the H ion concentration is of considerable importance in biological phenomena; for this reason it was thought worth while to arrange some experiments having for their object the determination of any changes in H ion concentration that may occur in a plant during its development.

In a paper by Clevenger (1) we find some values given for the H ion concentration of oats, buckwheat, soy-bean, and cow-pea. The determinations extended over a period of one month. The values of the H ion concentration at different times during the month cannot be compared conveniently, because in the earlier determinations whole tops were used together, while in the later ones the leaves and stems were determined separately. If the data obtained for the leaves are employed in the comparison, there is a decrease in actual acidity; on the other hand, the stems show a lower actual acidity than the whole tops. Taking the stems and leaves together and comparing the values thus obtained with those obtained for the whole tops, we find no appreciable change, except in the case of the oats, which show a slight decrease in H ion concentration.

Haas (2) reports that there is a change in H ion concentration of buckwheat plants from pH 5.48–5.97 in young plants to pH 4.82 in mature plants. According to these data, there is an increase in H ion concentration in buckwheat. The actual acidity of corn seedlings he finds also increases with the age of the seedlings.

In the present investigation three varieties of ordinary beans (Phaseolus) have been used: Robust pea bean, Great Northern and Crystal White wax bean. The plants were grown in six-inch pots, and after sprouting were thinned to only four plants to the pot. The three varieties were grown side by side. No attempt was made to control the acidity of the soil in any way. The plants were grown in the greenhouse during the winter; the experiments extended from November into April.

The plants were macerated (ground coarsely in a food chopper) and the juice expressed; only tops were used. This was done immediately after gathering the plants to avoid any change that might occur due to drying or other causes. The juice thus obtained was green, due to the chlorophyll, but not very viscous. It contained no large plant particles. The determination of the H ion concentration was made at once by a potentiometer.

An open electrode vessel was finally adopted and used in all experiments, after repeated failures with a closed Clark electrode vessel. All readings were taken five minutes after the electrode was immersed in the liquid. It was found by experience that this time was sufficient to allow an equilibrium to be established. In these determinations about fifteen cubic centimeters of liquid were used, and four or five different determinations were made one after another with different samples of juice from the same plants. These readings agreed with one another very closely. The apparatus was always checked against a known solution near the neutral point, before any experiments were performed, and frequently after the experiments. If there were any discrepancies between the two checks all data obtained were discarded.

Four sets (plantings) were carried out with each variety of bean. Graphs were constructed for each culture by plotting pH (H ion concentration) against age. The curves thus obtained were by no means identical, but they all showed the same general characteristics. There was always a greater H ion concentration in the young plants than in the mature plants. An average curve was constructed for each variety from the several cultures, by taking the average of them. These are shown in the accompanying graph (Fig 3).

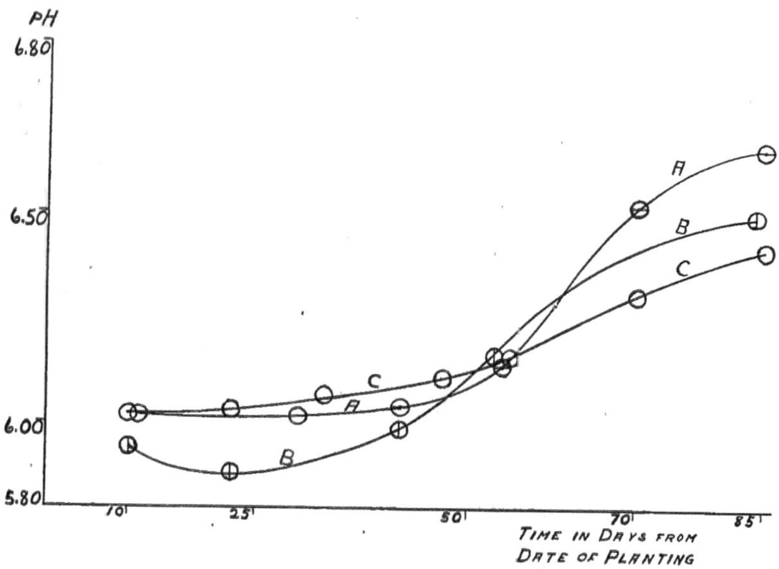

EXPLANATION OF GRAPH (FIG. 3)

Curve *A* represents the change in H ion concentration of Crystal White wax bean, and is the average of three different cultures. Curve *B* represents Great Northern and is the average of four different cultures. Curve *C* represents Robust pea bean, and is the average of three different cultures.

As shown by the curves, the H ion concentration of the three varieties is very nearly the same, at least in the early growth. The Great Northern variety showed an increase in actual acidity after ten days, which soon decreased. In the others there was a decrease from the start.

If we examine the graph we notice that there was only a small change during the first two months of growth. In the third month there was a much larger decrease in actual acidity. A change in appearance of the plants also became noticeable at this time. The lower leaves began to die; the earliest pods were maturing and only a few blossoms remained; no new growth was taking place.

The writer is not prepared to say to what this decrease in the H ion concentration is due. So far, only the fact that the change does occur has been investigated. It may be that, as the plant matures, proteins are formed whose reaction is neutral or even alkaline. This would neutralize the acids present and give a lower actual acidity. Another suggestion is that the substances (organic acids) giving the acid reaction in the young plants may be used up in the metabolism of the plant. Any number of other theories as to the cause of the change may be made, but they are all equally speculative.

SUMMARY

It has been shown, in three varieties of beans, that as the plants mature there is a decrease in the H ion concentration of the juice obtained from them.

UNIVERSITY OF MICHIGAN

REFERENCES

1. CLEVENGER, C. B. 1919. Hydrogen Ion Concentration of Plant Juices. Soil Science, 8:217.
2. HAAS, A. R. C. 1920. Studies on the Reaction of Plant Juices. Soil Science, 9:341.

THE GENUS ARMILLARIA IN THE UNITED STATES AND ITS RELATIONSHIPS

C. H. KAUFFMAN

The genus Armillaria is represented in this country by one very common species, *Armillaria mellea,* and by a number of other species, which by contrast to this, may, with few exceptions, be said to be remarkably infrequent or even rare; at least they are rarely collected and, except in local areas, never in large quantities. Dr. Peck in 1890 (17) gave a synopsis of the species at that time known from the United States. He cites the names of eight species: *A. mellea, A. nardosmia, A. ramentacea, A. constricta, A. robusta, A. bulbiger, A. mucida,* and *A. ponderosa.* He mentions three of these as occurring in New York State: *A. mellea, A. nardosmia,* and *A. ponderosa.*

In the *North American Flora,* Vol. 10, Part I, which appeared in 1914, Murrill (16) gives descriptions of fourteen American species. Of these, two species are tropical. Three species only, out of the eight mentioned by Peck, are included. The others are unmentioned, even in any synonomy.

In *Agaricaceae of Michigan* Kauffman (12) describes six species, two of which, *A. corticata* and *A. dryina* (See Plate V), are usually placed under Pleurotus by other authors. In his key, twelve species are given as possible for the eastern United States, some of which are based merely on their mention in American literature.

It appears, therefore, that the species of this genus are not only rare, but are very poorly known. With this in mind, the author feels that a critical account of the species now known to him from the growing condition in the field may aid in a more accurate evaluation of future collections and studies, and give the genus a more solid standing than it has recently received. It is to be noted that, in the latest extensive manual of the Agaricaceae by Carleton Rea (19), the genus Armillaria is left intact.

The group of Armillarias was first recognized by Fries (6) as a tribe under Agaricus. Quelet (18) raised it to the rank of a genus, a fact which has only historical value, since Fries clearly recognized its position and made the grouping of the species. Not all later mycologists have recognized this grouping, however, and the species belonging here according to the Fresian conception have at times all been placed under other genera, on the assumption that the genus Armillaria has no good grounds for existence. Thus Ricken (20), in *Die Blätter-pilze* has inserted seven European species, including *Armillaria mellea*, under the genus Clitocybe; nine species under Tricholoma; three species under Collybia; and four species under Pleurotus. Lange (14), in *Studies in Agarics of Denmark*, has likewise discarded the Armillarias as a genus, distributing them in the genera Tricholoma, Collybia, Pleurotus, and transferring *Armillaria mellea* to the genus Lepiota.

The essential characters of Armillaria as limited by Fries (9) are the white spores, the presence of a stipe which is continuous with the trama of the pileus, the attached gills, and especially the presence of a velum, which is at first innate with the stipe and the margin of the pileus; later, when this veil breaks down, the remnants decorate the stipe as a sheath or as scales, the upper portion sometimes terminating in a more or less flaring annulus. Although somewhat changed and emended in *Hymenomycetes Europaei* as compared with the wording in *Systema*, this has remained for most later mycologists the definite conception by which the species of this genus are recognizable.

The morphological development of a few species from the initial button stage has been examined by Robert Hartig (11), C. C. E. Fischer (7), Rudolf Beer (2) and George F. Atkinson (1). *Armillaria mellea* was studied by Hartig, Beer and Atkinson, and *Armillaria mucida* by Fischer. As a result of the work of the last three investigators we now know that in at least two species of Armillaria (and presumably this should be true in all of them), the hymenophore originates endogenously and is from the first enclosed by tissue which later represents the partial veil, the young fruit-body early showing further differentiation into the outermost layer usually referred to as the universal veil. This outer veil is more distinct below the margin of the pileus and downward on the stem than on the surface of the pileus; it apparently is best considered as only a partially differentiated cuticle of the pileus, and at least in its later condition is entirely analogous to the cuticle of the pilei in part of the genus Tricholoma and other similar genera or sections of these. In its manifestations on the stem of the maturing plants, this outer zone of tissue may, however, frequently differentiate sufficiently to be used as a morphological character in taxonomy, and has been so used in the genera Cortinarius, Tricholoma, Inocybe, etc. We are not yet certain how uniform the homologies of these veils are when applied to all the species of Armillaria; but from observations in the field I am satisfied that the outer (universal) and the inner (marginal) veils are developed very much the same in all species of true Armillarias at the stage when details can be made out with a pocket lens. There lie hidden in these characters, however, no broad lines of demarcation by which the species may be separated into several groups.

Few other contributions of any significance on the morphological structures of the species have been made since then. Fries himself had recognized the discreteness of the group by dividing it into sections on the basis of gill-attachment: those with adnate-emarginate gills, those with decurrent gills narrowed to a point on the stem, and those with equal, attached, non-emarginate gills and cartilaginous stems. These characters correspond with those of the genera Tricholoma, Clitocybe, and Collybia respectively. Ricken, using these same facts, argues that each group, therefore, is more closely connected with the genera mentioned and eliminates the genus Armillaria entirely.

Lange has proceeded along similar lines. He says (*l. c.*): "What is called 'the genus Armillaria' is properly speaking no genus at all, but a heterogeneous mixture of Agarics with white spores and a peronate or annulate stem. And the most satisfactory way of treating this spurious genus will therefore, I think, be to split it up altogether, distributing its several species among the adjoining genera."

It is admitted that there is some force to the statement that the genus Armillaria is a somewhat heterogeneous genus. It seems, however, that Lange has lost sight of the fact that in transferring *A. mellea* to Lepiota, he has only removed the difficulty to the genus Lepiota, which was already a somewhat heterogeneous genus itself.

Lange makes little of those traditional species of Lepiota, in which, as he himself points out, the gills are attached and the trama of the stipe is continuous with that of the pileus. These species, e. g., *L. amianthina, L. granosa, L. adnatifolia,* etc., are clearly aberrant Lepiotas, and should be removed. Although Fries and most mycologists since his time have retained the species of the "granulosae" section in Lepiota, Bresadola (3), commenting on the species *Armillaria Ambrosii* and *Armillaria haematites,* clearly intimates that that group of Lepiotas belongs more naturally to the genus Armillaria, by reason of their attached gills and the presence of a universal veil. It appears to me, then, that if any rearrangement involving the genus Lepiota is to be made, it would be more logical to transfer these species of Lepiota to the genus Armillaria rather than the reverse.

Superficially the view of Ricken and of Lange presents a good-looking phylogenetic scheme, since the highest development of the genus Tricholoma, for example, would thus be the annulate species, taken over from Armillaria, which had progressed to the point of differentiation represented by the veils. The same appears to apply to Clitocybe, Collybia and Pleurotus.

As Fayod (6) has shown in certain other genera, e. g., Hygrophorus, a microscopic examination of the gill-trama shows a structure quite constant, and by its correlation to other characters, very significant, of the natural relationships within that genus. In a transverse section of the gills, the hyphae lying between the hymenium or subhymenium are disposed either in a parallel, a divergent or an interwoven manner with reference to their orientation downwards from the pileus-trama to the edge of the gills; and these three modes of orientation are correlated with most of the other characters by which the genus Hygrophorus has been divided into three main groups. Kauffman (12) has verified these studies.

An examination of the gill-structure of the species of Armillaria reveals in part the same differences that are found in species of Hygrophorus, representing at least two such hyphal orientations. The larger number studied have a parallel gill-trama; this in turn is often correlated with adnexed or emarginate gills. Both the parallel hyphae of the gill-trama and the adnate-emarginate mode of gill-attachment are Tricholoma characters. Such Armillarias, however, possess usually both a well-developed universal veil and a partial veil, and thus could of course be segregated, as was done by Ricken, as a small group which terminate the Tricholoma line of differentiation. Fries, in his *Monographia*, recognizes this fact, and the effort of Ricken and Lange to reduce this Armillaria group to Tricholoma standing does not add any new viewpoint.

Using the same methods we find that certain Armillarias with well-developed veils, attach themselves in the same way to the genera Collybia, Pleurotus, and, as already pointed out, to the genus Lepiota. In these groups, also, as far as I have data, the gill-hyphae are disposed in a parallel fashion.

Certain species of Armillaria, however, have decidedly decurrent gills. In the known species of this kind, the plants are usually large and compact, and the veils are well developed, forming a sheath on the stem, and under favorable conditions for development terminating in a thick, flaring annulus. Fries considered these as topping the line represented by the genus Clitocybe, and Ricken has joined them to that genus. From an examination of the gill-structure of the species of this group so far available, it would appear that here the gill-hyphae are not parallel, but *diverging.* A considerable number of species of the genus Clitocybe which I have studied in this respect possess always parallel gill-hyphae, and I suspect such is the rule for all of them. This situation, therefore, militates against treating the decurrent-gilled Armillarias as the topmost branch of the genus Clitocybe, as has always been done. We must seek a connecting line somewhere else. These Armillarias, furthermore, do not have the general aspects of the larger species of Clitocybe, but rather of some of the Tricholomas.

Divergent gill-hyphae occur in Amanita, in the subgenus Limacium of the Hygrophori, and in the viscid Lepiotas. In fact the presence of gelatinizing layers on the pileus, or as the outer portion of universal veils, seems to be largely correlated with this particular gill structure. It seems impossible, however, to push aside the other generic characters of Amanita, Hygrophorus or Lepiota and to associate with them the decurrent-gilled species of Armillaria.

In so far, then, as the decurrent-gilled species are concerned, we have a good morphological foundation for retaining the genus Armillaria. As to the other three groups, it seems an open question whether they should be segregated as three distinct genera, each the top-piece of the genera Tricholoma, Collybia, Lepiota and Pleurotus respectively, or whether they should be put together into the tribe Armillaria, distinguishable by the common bond of the well-developed veils. The latter scheme seems to have in its favor both a regard for their traditional disposition and practical taxonomic convenience, entirely apart from the question of ultimate relationships; and since Fries, when he established the genus Armillaria, clearly pointed out this polyphyletic relationship, there seems to be no point in breaking up the genus now.

Unfortunately, of the ten species with decurrent gills, which, according to the literature, might be expected to occur in the United States, only three have been carefully studied to show that they have a divergent gill-trama. It is desirable that those who may run across these rare species, should turn their attention to this character, as well as to the histological and developmental phases of the veils.

Synopsis of the Species of Armillaria of North Temperate Regions

(Compiled in part from the literature)

1. Gills decurrent.. (2)
1. Gills adnate to emarginate or uncinate.......................... (11)
2. Growing on rotten wood, from old roots, stumps, trunks, etc...... (3)
2. Growing on the ground.. (6)
3. Stems eccentric or lateral; pileus white or whitish............. (4)
3. Stems usually central.. (5)
4. Spores 9–10 × 4–4.5 μ............................*A. dryina* Fr.-Pat.
4. Spores 13–17 × 4–5 μ..........................*A. corticata* Fr.-Pat.
5. Very common; honey-yellow, especially at base of stem, becoming darker in age; pileus dry; stems caespitose, attached to rhizomorphs..............(*A. putrida* (Scop.) Murr.) *A. mellea* Fr.
5. Pileus 5–8 cm. broad, white, with brownish-scaly disk, *dry*; annulus thin, white, membranous; stem solid, brown-scaly-dotted below the ring. (Spores unknown.).........*A. rhagadiosa* Fr.-Ricken (Peck considered *A. nardosmia* a segregate of this. See Farlow (5), *Index*, p. 257.)
5. Pileus 3–5 cm. broad, brown-gray, *viscid;* especially on mulberry trees. (Spores unknown.)...................*A. mori* Paul-Ricken
6. Stem hollow.. (7)
6. Stem solid.. (8)
7. Pileus viscid, 5–6 cm., white, brown center, *striate to middle;* under conifers; annulus dependent. (Spores unknown.) *A. subconcava* Schum.-Ricken
7. Pileus dry, 8–13 cm. broad, white at first to creamy (rich ochre with reddish tint when dried); veil large, white, forming a thick, flaring annulus, sheathing the stem below; *gills* very broad in front, attenuate at stem, subdistant. (Spores 14–17.5 × 3–5 μ.) (*Catathelasma evanescens* Lovejoy) *A. evanescens* (Lovejoy) Murr.
8. Annulus narrow, membranous, rather thin; stem dry, solid, with slight fibrillose shreds; pileus 7–10 cm., sordid-alutaceous, pallescens, glabrous; gills very attenuate at both ends, long-decurrent, crowded. (Spores unknown.).............*A. Laschii* Fr.
8. Annulus thick, terminating the well-developed sheathing veil, at least in vigorous plants.................................. (9)
9. Pileus fuscous-brown, 8–15 cm. broad, with darker appressed scales toward margin; gills narrow, crowded; annulus ample, persistent; stem scaly below annulus. Under conifers. (Spores 12–14 × 5–6 μ Murr. Sacc.) (Spores 11–13 × 5–6 Peck). (*A. nobilis* Murr.) *A. imperialis* Fr.
9. Pileus and stem white or whitish.......................... (10)
10. Spores 10–12 × 5–6 μ; pileus and stem not viscid, white or nearly so.......................(*Lentinus ventricosa* Pk.) *A. ventricosa* Pk.

10. Spores 14–16 (17) × 5–6 (7) μ; pileus and stem subviscid from the outer thin pellicle of the universal veil, "pinkish-buff" to "cinnamon-buff" (Ridg.) .... *A. macrospora* Pk.
11. Growing caespitose or subcaespitose on stumps or logs, trunks, branches, or cut timber; mostly on long stems .... (12)
11. Growing on the ground, or among mosses, humus or débris .... (14)
12. Pileus granulose, rugose-wrinkled, ochraceous to fulvus; stem floccose-peronate, with flaring annulus; spores 4–5 × 3 μ (*Lepiota granosa* (Morg.) *A. granosa* (Morg.) comb. nov.
12. Pileus not granulose .... (13)
13. Pileus white, viscid; on dead *beech* and *birch* trunks or branches; spores subglobose, 14–18 μ. (Ricken: 15–18 μ.) (Stevenson: 14–17 μ) .... *A. mucida* Fr
13. Pileus tawny to chestnut brown; on or around conifer wood; gills free; pileus subviscid; stems fuscous-scaly below annulus; odor heavy, of rancid meal. Spores globose 5–6 μ. *A. megalopus* Bres.
14. Pileus not granulose-warty .... (18)
14. Pileus granulose or granulose-warty; plants Lepiota-like .... (15)
15. Gills adnate; stems rather long and slender .... (16)
15. Gills adnexed; stems short .... (17)
16. Spores 6–7 × 3.5 μ (Lange); cystidia none; pileus usually umbonate, reddish-ochraceous. (*Lepiota amianthina*) *A. amianthina* (Fr.) comb. nov.
16. Spores 5–5.5 × 2.5–3μ (Kauffman); cystidia present but scattered; pileus rusty-red. (*Lepiota adnatifolia*) *A. adnatifolia* (Pk.) comb. nov.
17. Pileus varying "ochraceous buff" to "cinnamon-rufous" (Ridg.) (*Lepiota granulosa*) *A. granulosa* (Fr.) comb. nov.
17. Pileus varying "ferruginous" to "English-red (Ridg.). (*Lepiota granulosa cinnabarina*) *A. cinnabarina* (Fr.) comb. nov.
18. Pileus white, creamy white to buff, sometimes sordid ochraceous or rust-colored in age, not with markedly uniformly yellow shades or red shades .... (19)
18. Pileus either at first with yellow or rufous colors, or soon becoming so .... (24)
19. Pileus 3–5 cm. broad, dry .... (20)
19. Pileus usually much larger .... (21)
20. Odor strongly of fresh meal; stem solid; all parts white; spores 7–8 × 4–5 μ, elliptical (Ricken); in grassy places under willows, etc. .... *A. constricta* Fr.
20. Odor none; stem equal, stuffed-hollow; pileus, etc., at first snow-white, then yellowish on disk when old; under Larix; spores elliptical, 4–5 × 2.5 μ (Bres.) .... *A. Ambrosii* Bres.
21. Veil (probably outer layer) subviscid, at least on stem; stout plants .... (22)
21. Veil, etc., dry, gills rather broad; stout plants .... (23)
22. Spores ellipsoid, 5–6 × 3–3.5 (4) μ (Kauffman), odor penetrating, subalkaline; gills narrowed behind, at first slightly subdecurrent-acuminate .... *A. viscidipes* Pk.

22. Spores globose or subglobose, about 4 μ (Peck); gills narrow, subemarginate; veil prominent, sheathing; odor and taste mild..............(*A. magnivelaris* Pk.) *A. ponderosa* (Pk.) Sacc.
23. Spores ellipsoid, 7.5 × 5 μ (Peck); stem bulbous, tapering upwards; described from Alabama..............*A. appendiculata* Pk.
23. Spores globose or subglobose, 5–6 × 4.5–5 μ (Kauffman); stem tapering downwards or subequal; from Oregon and Washington..*A. arenicola* Murr.
24. Reddish or brown shades predominating.................... (25)
24. Yellow predominating.................................... (28)
25. Like *Lepiota amianthina* in habit, etc., but color incarnate-vinose; annulus cortiniform; stem fistulose...*A. haematites* B. & Br.
25. Not with these characters................................ (26)
26. Plants firm, compact; stems stout, short-pointed at base...... (27)
26. Plant of rather soft texture; pileus and stem tile-red or cinnabar-incarnate; odor farinaceous; spores 4–5 × 3 μ (Ricken). *A. focalis* Fr.
27. Pileus rufescent, soon rimulose-diffracted; gills relatively broad, becoming rufous-spotted, adnate; odor subfarinaceous; spores 4–5 × 2.5–3 μ (Kauffman) . . . .*A. robusta* Fr.
27. Pileus brownish, spotted with thin, appressed rufous-brown to blackish-brown scales; gills medium broad, white, sinuate-adnate; odor slight or fruity; spores 6–7.5 × 5 μ (Kauffman) (? *A. nardosmia* (Ell.) Sacc.) *A. caligata* Fr.
28. Spores globose, 4–5 μ (Ricken); stem solid; pileus dry, straw-yellow, tinged greenish in age; gills deeply emarginate. *A. luteovirens* Fr.
28. Spores ellipsoid, 6–9 × 4–5 μ (Atkinson), 4–5 (6) × 3–3.5 μ (Kauffman); stem stuffed-hollow; pileus viscid, "apricot-yellow" (Ridg.), tinged tawny in age; gills adnate, rounded behind...................................*A. albolanaripes* Atk.

COMMENTS

Armillaria evanescens and Armillaria macrospora (See Plate VI) were described from Wyoming and Colorado respectively. Both species have decurrent gills and strikingly large spores. I have collected such plants in Colorado, and am strongly of the opinion that these Rocky Mountain species are identical. The spores easily vary to the extent of the difference recorded for each species, and the development of the universal veil on the stem is more or less vigorous according to conditions. I have given a full revised description of *A. macrospora* in a previous paper (13).

Armillaria ventricosa. — First described by Peck as a Lentinus from the District of Columbia. It was later changed by him to its present position. It also has been reported once each from Alabama, North Carolina and Nova Scotia. The northern collection may well be considered as a doubtful determination. An excellent photograph, taken by H. S. Beardslee, of the fresh plant collected by him in North Carolina, was reproduced in the *New York State Mus. Bull.*, Nos. 205–6, Fig. 3, 1918.

Armillaria mucida. — American notices on this species are vague and unsatisfactory. It is reported from Pennsylvania southward in the coast states by McIlvaine, Schweinitz and Curtis, but apparently no authentic specimens are in existence. It is quite frequent in England, and will probably be rediscovered in the Eastern United States. Good illustrations of it are given by C. E. C. Fischer (*l. c.*).

Armillaria constricta. — This is one of the smaller species, and was reported from California by Harkness. Its status as an American plant is not very good.

Armillaria viscidipes. — There seems to be only one collection of this species known in the East: Dr. Peck collected it in Dutchess County, New York, in 1890. I have already reported its occurrence in Colorado (13). The western form has been studied carefully, and except for the fact that the spores average slightly smaller than the size given by Peck, it seems impossible to consider it distinct. I found considerable difficulty to determine accurately the spore size of the different collections made, as the spores often mature slowly; many seem globular in appearance when young (although this is often due to the more prominent oil-globule), and in some fully developed plants the spores are somewhat smaller than in others. Under the oil-immersion lens, they are delicately rough-punctate.

Special attention was directed to the gills. In the young plants they are at first adnate, then gradually become arcuate and slightly acuminate-decurrent; in old plants they are sinuate near the stem. Peck in his description says the gills of his plants were "sinuate or subdecurrent," but in his figures

(*N. Y. State Mus. Bull.*, 44: Pl. II), he shows them distinctly decurrent, a condition not at all indicated by his description, and not present in the Colorado form.

It is to be noted also that Peck does not mention any viscidity on the pileus. In rather dry weather the western form also would be considered as lacking viscidity, but plants collected in moist places or after rains not only had a distinct viscid layer on the surface of the stem where the latter was coated by the universal veil, but also a very thin viscid layer on the surface of the cap, composed as usual of a gelatinizing hyphal coating. I strongly suspect this will be the case in the eastern form when it is again found and carefully studied. The striking common characteristic of both forms is the rather strong, penetrating subalkaline odor. If these reasons for the identity of the western and eastern plants are correct, it should be desirable to have a description of the Colorado form, as follows:

Pileus 6–12 (15) cm. broad, fleshy, firm, convex or campanulate-convex, obtuse, then expanded to plane, margin at first incurved and tomentose, subviscid when moist, sometimes shining when dry, whitish or "light ochraceous-buff" (Ridg.) when young and fresh, becoming stained "ochraceous-tawny" or varying with ochraceous or tawny shades or stains or streaks, glabrous, even, sometimes rimulose in age; flesh compact, thick, abruptly thin at margin, white. Gills adnate to slightly acuminate-subdecurrent, at length sinuate, crowded, rather narrow, whitish, sometimes ochraceous-stained near edge, trama of parallel hyphae. Stem stout, 4–8 (10) cm. long, 2–3 cm. thick, equal or mostly tapering at base, solid, firm, sheathed at first by the rather thin, membranaceous, "ochraceous-buff," universal veil, which at length is terminated by a more or less flaring, narrow, membranous annulus; sometimes the stem is merely marked by thin concentric remnants of the broken veil, the remnants becoming ochraceous-tawny like the pileus, white and flocculose above the annulus. Spores 5–6 × 3–3.5 (4) μ, short-ellipsoid to oval, smooth, often with prominent oil globule. Odor usually strong, penetrating, subalkaline, persistent. Distribution: New York, Colorado, Oregon.

ARMILLARIA PONDEROSA and ARMILLARIA ARENICOLA. — These two species seem very close, the former being recorded from New England and New York, the latter from Oregon. I have collected at Hoodsport, Washington, what appears to be a rather common species in certain regions of Washington, common enough to be assiduously collected by the Japanese for commercial purposes. I have referred this collection to *A. arenicola* (See Plate VII), but except for the recorded account of the spores of *A. ponderosa*, it could be just as well referred to the latter. The odor and taste of the western plants, which are edible, are mild.

ARMILLARIA ROBUSTA (See Plate VIII). — My first collection of this interesting European species was made at Seventh Lake, Adirondack Mountains, New York, under some white birches at the summer cottage of Prof. F. C. Stewart. The fresh young plant when cut has a strong farinaceous odor, which, however, is less marked in older plants; the taste is bitterish-farinaceous. The plants grow rather squat. The stems are short, 3–5 (7) cm. long by (10) 15–20 mm. thick, subequal or tapering downwards, solid, sheathed by the thin fibrillose universal veil which terminates above in a floccose-fibrillose, narrow and subevanescent annulus. The firm pileus is relatively broad, 5–10 cm., with a compact, white context; its dry surface, although at first glabrous and even, is soon rimose or areolate, sometimes diffracted-scaly. The most striking character is its rufescent tendency. The young pileus and stem are "hazel"-colored, but this soon changes to "rufous" or "cinnamon rufous" (Ridg.). The gills are at first adnate, with decurrent lines, soon deeply emarginate-adnexed, broad, crowded and whitish, but like the other parts they become rufescent-spotted. The spores measure $4–5 \times 2.5–3\ \mu$, and the gill-trama is composed of parallel hyphae.

It has been reported from New England southward to the Carolinas (15). It is not easy to determine, however, to what extent these reports are authentic.

ARMILLARIA CALIGATA. — A full discussion of this species appeared in *Agar. Mich.*, 1: 650 (12).

ARMILLARIA ALBOLANARIPES (See Plate IX). — This western species has been found in Washington, Oregon and California.

I have collected it in the Olympic Mountains. The pileus is 5–10 cm. broad, viscid, "apricot-yellow" (Ridg.) when fresh, with "tawny" centre and sometimes with obscure, appressed, tawny scales; flesh white except under the cuticle, and soon softer than in other species. The gills are adnate, rounded behind, broad, close to subdistant, thin, whitish and at length tinged lemon-yellow. The stem is 5–8 cm. long, 8–12 mm. thick, and at first sheathed up to the lacerate-floccose yellow annulus by more or less tufted, or concentric, soft, white, floccose scales, which may disappear in age, white within and at first stuffed by a soft pith. The odor and taste are mild. It is gregarious to subcaespitose in fir or hemlock forests. The spores measure 4–5 (6) × 3–4 μ; the larger sizes given by Atkinson are rare in a mount. It appears in the *North American Flora* as *A. albolanatipes.*

EXOTIC, EXCLUDED OR DOUBTFUL SPECIES

ARMILLARIA ALPHITOPHILA (B. & C.) Murr. — Reported in the *North American Flora* from Louisiana, Mexico and the West Indies.

ARMILLARIA AURANTIA Quèl. — Usually considered as a Tricholoma. *Tricholoma Peckii* Howe is a synonym.

ARMILLARIA BULBIGERA Fr. — The consensus of modern opinion places this species in the genus Cortinarius, in spite of the statement by Fries (*Icones* 1: Text, p. 23) that "it is in no way related" (*Nullo modo affinis*). I have not become acquainted with the growing plant if indeed it occurs in the United States.

ARMILLARIA CHEIMONOPHYLLUS B. & C. — Placed in synonomy under *A. alphitophylla* in the *North American Flora.* Not an Armillaria, according to Farlow (*Bibl. Index, Vol.* 1, p. 252).

ARMILLARIA CITRI Inzengi. — Illustrated by Cooke (4) (Pl. 1181); probably introduced at Kew Gardens, and exotic.

ARMILLARIA COLOSSA (Fr.) Boud. — As shown by Boudier (*Bull. Soc. Myc.*, 16: p. 18, Plate I), this species, usually referred to *Tricholoma,* is a good Armillaria.

ARMILLARIA FRACTICA (Britz.) Sacc. — Pileus large, glutinous, yellow to reddish-brown; gills adnexed; stem whitish, with the annulus colored like the pileus; odor farinaceous; spores globose, 4–5 μ; from Bavaria.

ARMILLARIA JASONIS Cke. & Massee. — Habit of a Pholiota, but spores white; on stumps. England. Well illustrated by Cooke (4) (Pl. 955).

ARMILLARIA RAMENTACEA Fr. — The position of this species is not very clear. It seems to be close to *Tricholoma terreum*, *T. argyraceum*, etc.

ARMILLARIA ROBOROSA Britz. — Spores "not hyaline," according to its author. Not an Armillaria.

ARMILLARIA SOLIDIPES Pk. — This species was imperfectly described and has been referred by Murrill to *A. mellea* as a synonym.

ARMILLARIA SQUAMEA Barla. — From the Maritime Alps. Nothing further known.

ARMILLARIA SUBANNULATA Pk. — This species, from the Pacific Coast, is, according to Murrill and Zeller, a Tricholoma. Zeller, in *Mycologia*, XIV. 187, gives additional data.

ARMILLARIA SUBDEHISCENS (Britz.) Sacc. — Pileus glutinous, flame-yellow to red-brown; spores 6 × 4–5 μ; gills emarginate.

ARMILLARIA SUBIMPERIALIS (Britz.) Sacc. — Spores 12–14 × 5–6 μ. Bavaria. Probably *A. imperialis* Fr.

ARMILLARIA UMBILICATA Pat. — Reported in the *North American Flora* from Guadaloupe.

ARMILLARIA UMBONATA (Sumstine) Murr. — Originally described as an Amanita. See *North American Flora* for description.

UNIVERSITY OF MICHIGAN

LITERATURE CITED

1. ATKINSON, G. F. 1914. The Development of Armillaria mellea. Myc. Centrbl., 4: 113.

2. BEER, R. 1911. Notes on the Development of the Carpophore of Some Agaricaceae. Ann. Bot., 25: 683.

PLATE V

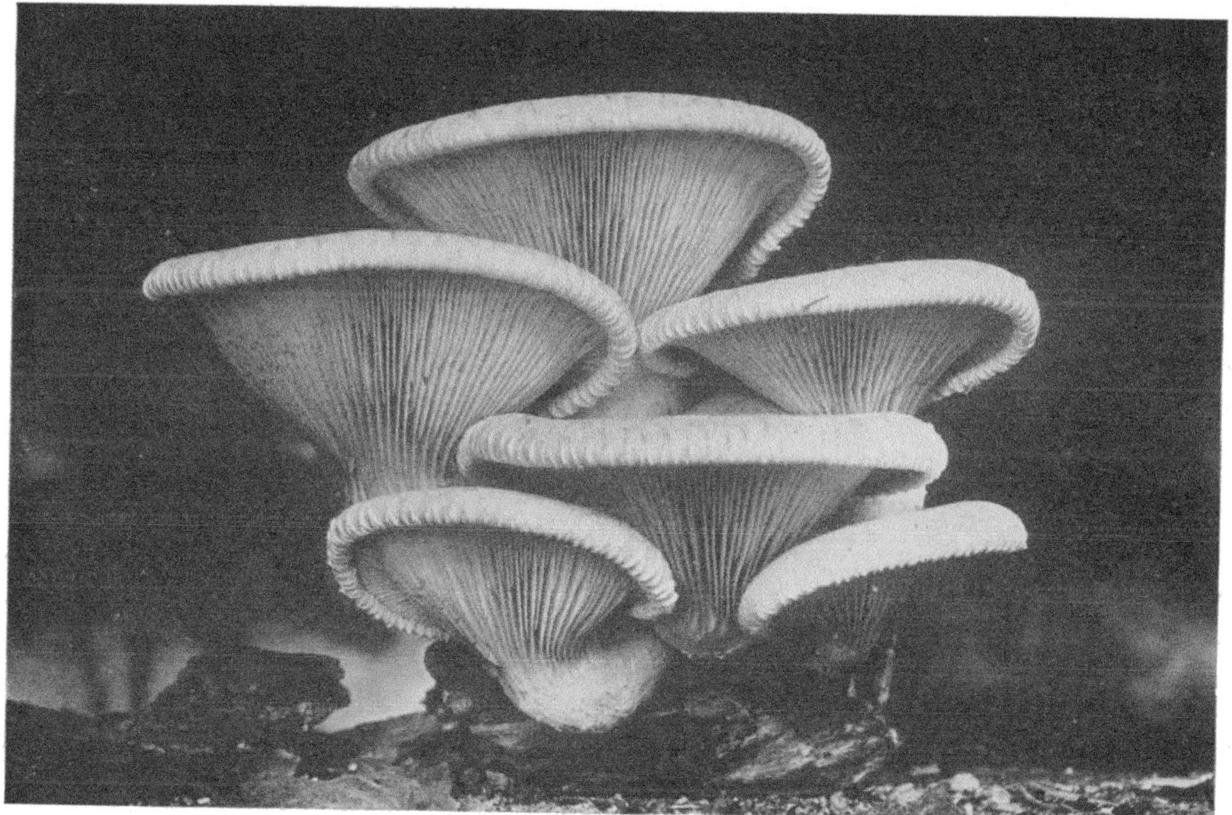

Armillaria dryina

PLATE VI

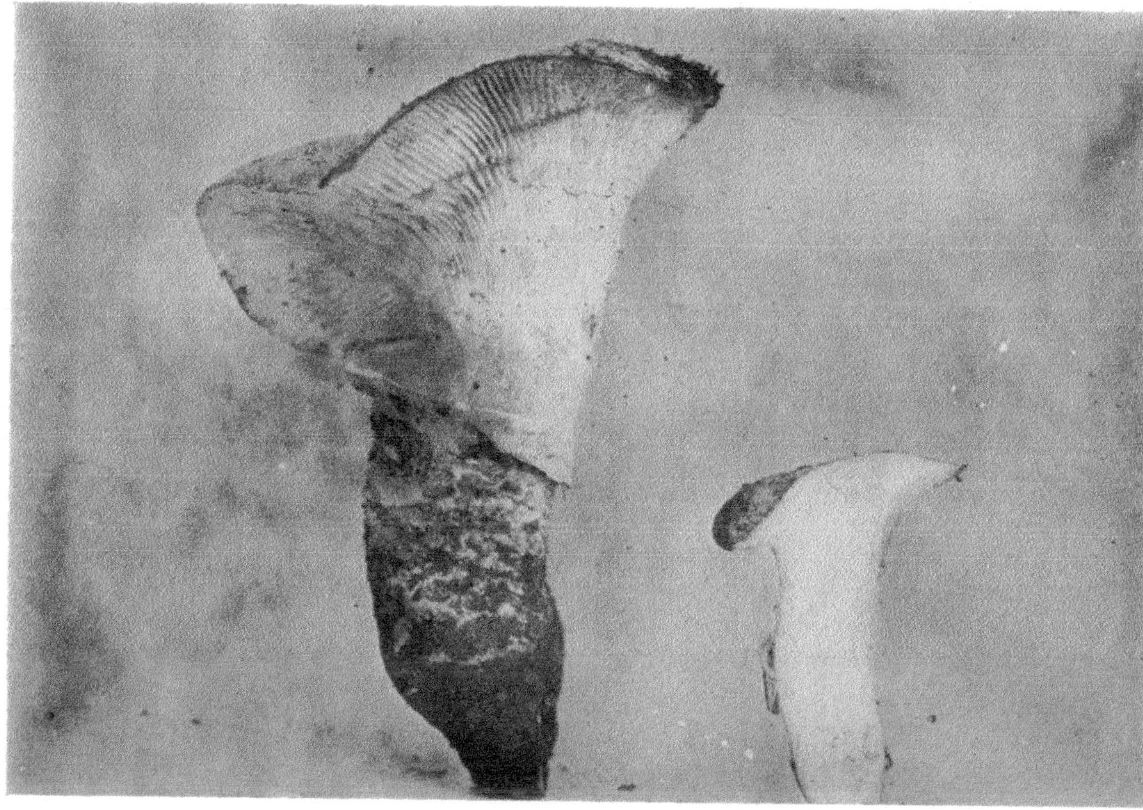

ARMILLARIA MACROSPORA

PLATE VII

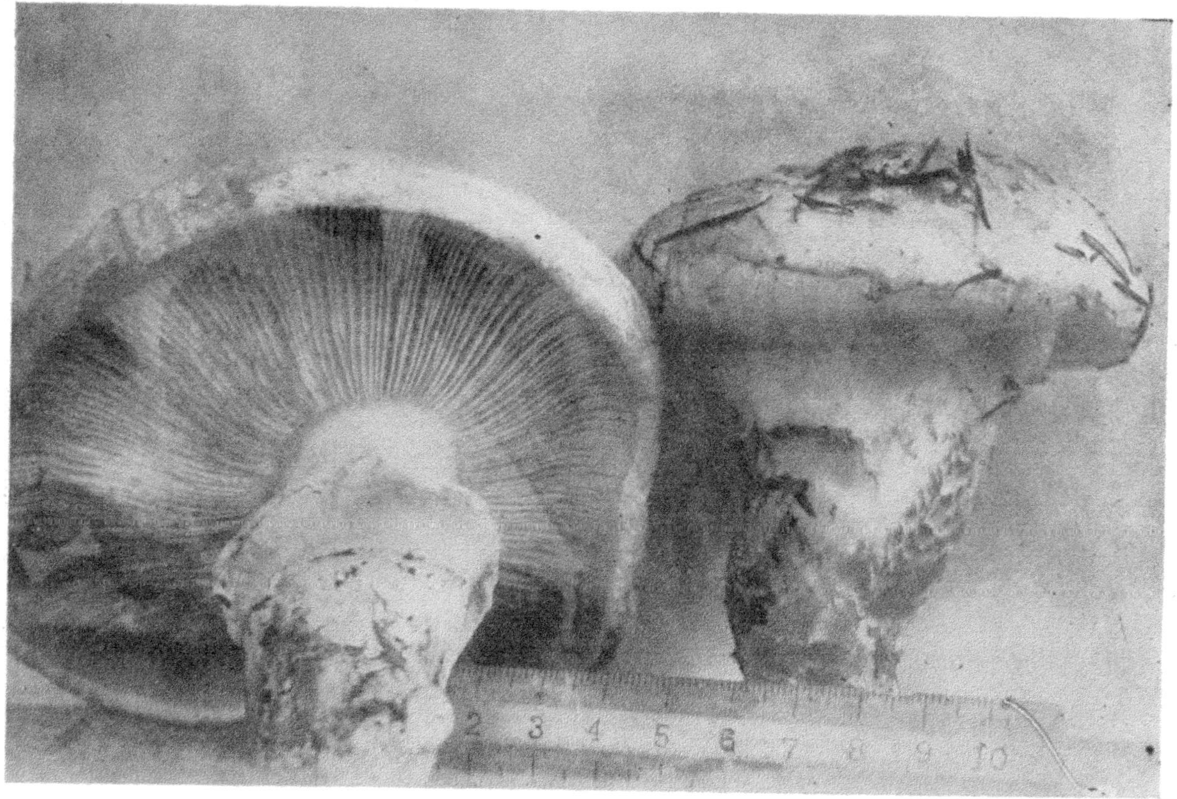

Armillaria arenicola

PLATE VIII

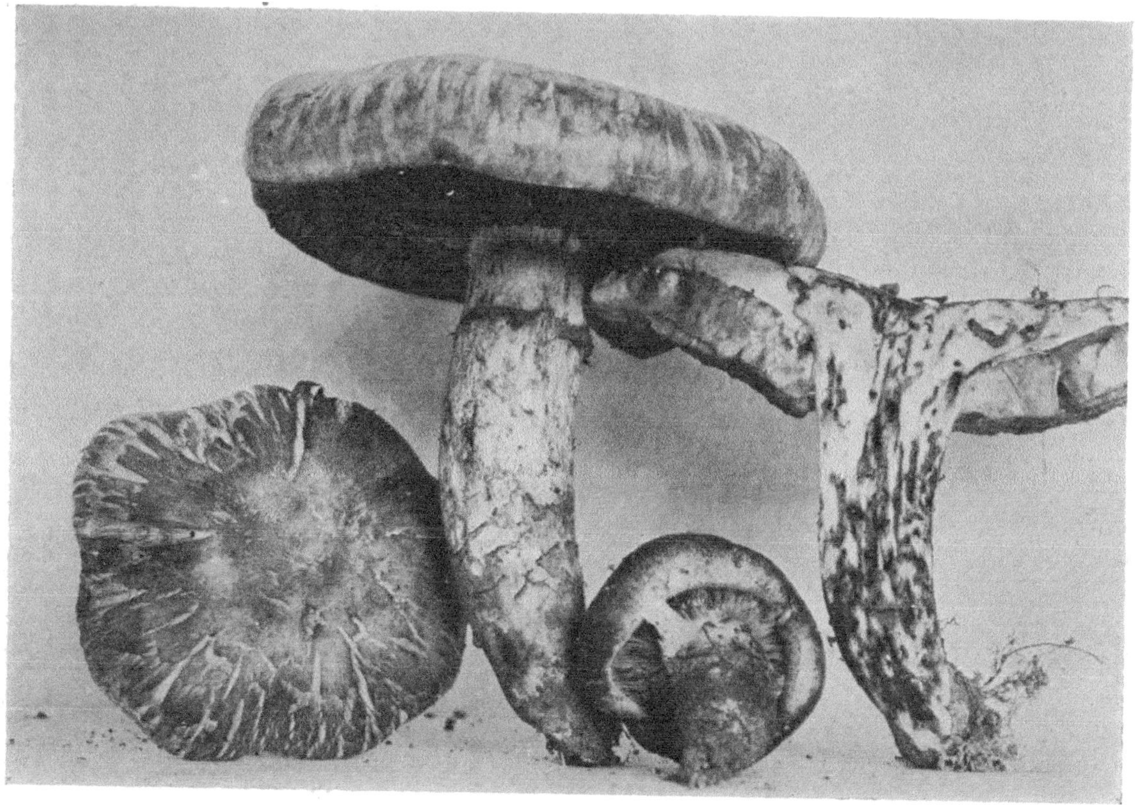

Armillaria robusta

PLATE IX

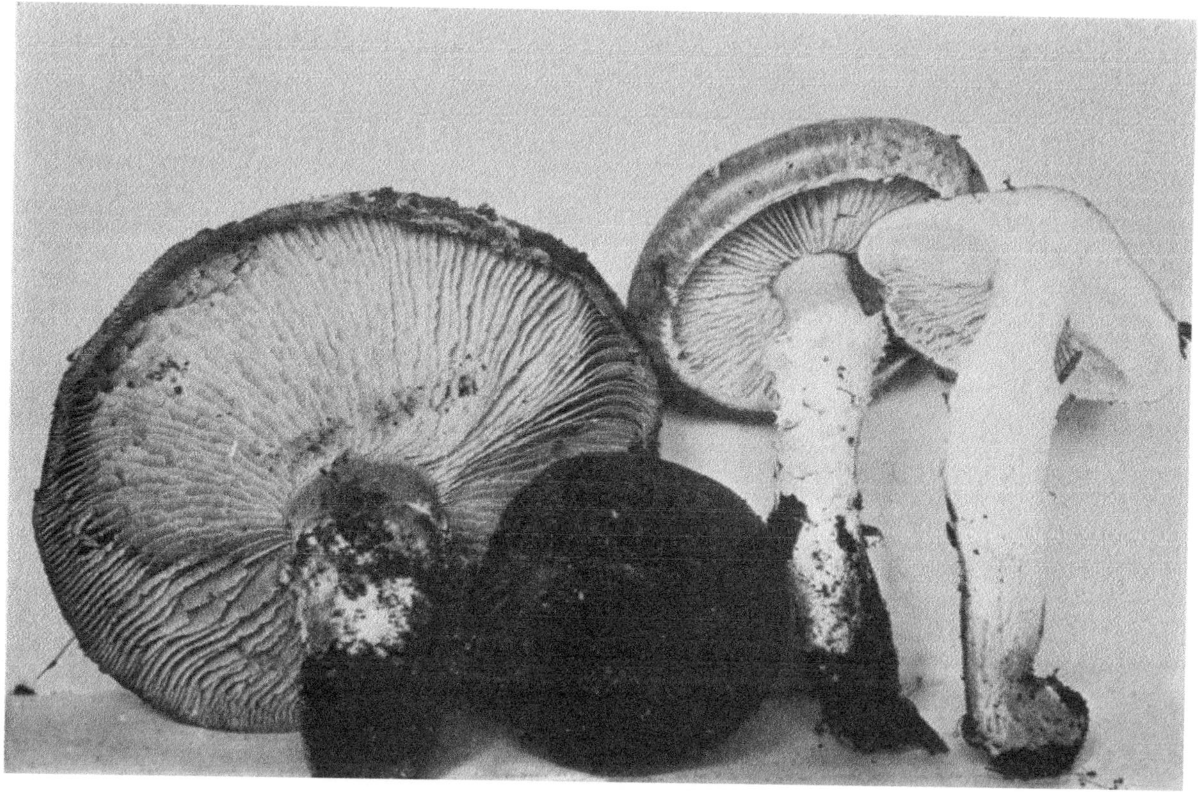

Armillaria albolanaribes

3. Bresadola, G. 1881, 1892. Fungi Tridentini, I: 27. 1881; II: 5. 1892.

4. Cooke, M. C. 1865–74. Illustrations of British Fungi.

5. Farlow, W. G. 1905. Bibliographical Index I, Part I: 252.

6. Fayod, V. M. 1889. Prodrome d'une Histoire Naturelle des Agaricinées. Ann. Sci. Nat. Bot., 7th ser., 9: 181.

7. Fischer, C. E. C. 1909. On the Development of the Fructification of Armillaria mucida (Schrad). Ann. Bot., 23: 503.

8. Fries, Elias Magnus 1821. Syst. Myc., I: 26.

9. —— 1874. Hymen. Europ., p. 40.

10. —— 1857. Monographia, I: 34.

11. Hartig, R. 1874. Wichtige Krankheiten der Waldbäume. Berlin.

12. Kauffman, C. H. 1918. Agaricaceae of Michigan. I: 647.

13. —— The Mycological Flora of the Higher Rockies of Colorado. 1921. Papers Mich. Acad. Sci. Arts and Letters, 1: 100–150. (Published 1923.)

14. Lange, Jakob E. 1915. Studies in the Agarics of Denmark, Part II. Dansk Botanisk Arkiv., 2: 13.

15. McIlvaine & Macadam. 1912. One Thousand American Fungi. Revis. ed., 52.

16. Murrill, W. A. 1914. North American Flora, 10: 36.

17. Peck, C. H. 1890. N. Y. State Mus. Rep., 43: 86.

18. Quelet, L. 1872. Les Champignons du Jura et des Vosges, p. 36.

19. Rea, Carleton. 1922. British Basidiomycetes. Cambridge.

20. Ricken, A. 1914. Die Blätterpilze Deutschlands, I: 329; 361; 400.

TWO UNREPORTED PARASITES OF *HEVEA BRASILIENSIS*

CARL D. LA RUE

Since the Para rubber tree, *Hevea brasiliensis* Muell.-Arg. was brought to the East Indies from Brazil, it has been attacked by many of the large number of organisms already parasitic on a miscellaneous group of host plants in that habitat. The occurrence of a considerable number of these parasites on *Hevea brasiliensis* has been recorded; the fulness of the account of the attack is generally in direct proportion to the economic effect of the depredations of the organism. The ravages of the two parasites discussed here have been relatively limited in extent. Consequently they have thus far escaped discussion and they are now reported more by reason of their interest than because of their importance.

SCLEROTIUM SP. — The latex from which rubber is secured is taken from the Para rubber tree by cutting away the outer bark of the tree, a process that leaves only about 2 mm. of young bark outside the cambium. Each day a shaving of bark extending from the surface of the tree to within 2 mm. of the cambium and about 2 mm. thick is cut away to open the tubes from which the latex flows. By this method, which is called "tapping," the bark is continuously removed, and a layer of young, tender cells immediately outside the cambium is exposed. In a short time a layer of cork is formed which covers this young bark, but before this is developed the exposed cells are subject to attack by a number of parasites, among which a species of Sclerotium was found by the author on trees in Sumatra.

The mycelium of the fungus spreads over the exposed surfaces and quickly kills the bark which has been tapped within two months of the time of the attack, but bark tapped prior to this time seems to be sufficiently protected by the corky layer to resist infection. The earliest infection on a tree usually takes place on bark which has been very recently tapped. This is to be expected as in this region there are many cut and bruised cells which supply food to the developing mycelium. The partially protected bark which has had a month or more to recover from the effect of tapping can be parasitized only by a mycelium which is well established.

The formation of sclerotia begins as soon as the mycelium is well grown. This occurs equally well in the field and on pieces of bark which have been removed from the trees and kept in moist chambers in the laboratory.

The fungus resembles *Sclerotium Rolfsii,* especially in regard to the size and color of the sclerotia, but as descriptions of the many species of Sclerotium were lacking in Sumatra where these observations were made the species remains undetermined.

Cephaleuros virescens Kunze. — This is a common parasite on *Hevea brasiliensis* on the East Coast of Sumatra where, however, its attacks are noticeable only on trees which, for any reason, are lacking in vigor. On the twigs the parasite appears as a reddish brown scurf which may cover the greater part of the surface. It appears to cause a stunting of the twigs on which it grows, as such twigs are noticeably less vigorous than those on which the parasite is not found. The infected twigs, however, were undoubtedly already weak before the invasion of the parasite, which appears to be able to attack seriously none but weakened plants. The orange-yellow fruiting bodies of the alga are rarely found on twigs of Hevea but are common on twigs of plants such as *Graptophyllum pictum* on which they are sometimes so numerous as to cover the twigs almost completely. No other plant has been seen on which the alga fruits so freely.

The organism occurs on leaves of Hevea much more infrequently than on the twigs. It does not develop to any extent on the exterior of the leaf, but spreads in the interior, injuring and discoloring the mesophyll for a distance of from 1 to 3 mm. around the point of infection. From these infected areas the fruiting bodies push out to the surface and frequently their appearance is the only external indication of the existence of the infection.

The alga appears to have spread to the rubber from some of its numerous hosts in the jungle or in the native villages. The author has observed it on the mango, the horse mango, the Bougainvillea, and the common shrub, *Graptophyllum pictum* (L.) Griff. In British India it has been found on a number of hosts by Cunningham,[1] by Ward,[2] and by Mann and Hutchinson.[3] The "red rust" disease of tea is caused by this organism, according to Mann and Hutchinson, who state that its attacks are serious only on weakened plants, as has already been stated to be the case with rubber trees. The writer has observed that while normal plants of *Graptophyllum pictum* are parasitized to a relatively slight degree, one group of plants on an area of barren, sandy soil was very severely attacked. The surfaces of the stems were almost completely covered with fruiting bodies and the leaves were badly spotted. It was also noticed that the strong shoots with green leaves were less subject to attack than were the weaker ones with variegated leaves.

It is not believed that this parasite has a very considerable effect on either the growth or the yield of the rubber trees, but its attacks may be of consequence under certain conditions. When a tree is to be used as a source of scions for grafting or budding, the twigs must be in a vigorous condition; otherwise the cambium will be too inactive to unite readily with that of the stock. In his work on the vegetative propagation of rubber trees the writer has encountered scores of trees with high rubber yields to their credit, which would give desirable buds were they only in better growing condition. The twigs of such trees are generally covered with patches infected by this alga, and there is no doubt that it is a factor which contributes to, if it does not cause, this weakened condition of the trees.

University of Michigan

[1] Cunningham, D. D. *On Mycoidea parasitica, a new genus of Parasitic Algae, and the part which it plays in the formation of certain lichens. Trans. Linn. Soc. London, Sec. Series, Bot.* 1:301–316. 1880.

[2] Ward, H. Marshall. *On the Structure, Development, and Life-history of a Tropical Epiphyllous Lichen* (*Strigula complanata*, Fée, fide Rev. J. M. Crombie). *Trans. Linn. Soc. London, Sec. Series, Bot.* 2:87–119. 1881–87.

[3] Mann, H. H. and Hutchinson, C. M. *Cephaleuros virescens*, Kunze, *The Red Rust of Tea. Memoirs of the Dept. of Agric. in India.* (*Botanical Series*), 1. No. 6. 1907.

A LEAF-FALL DISEASE OF *HEVEA BRASILIENSIS* MUELL.-ARG. DUE TO *GLOEOSPORIUM ALBORUBRUM* PETCH*

CARL D. LA RUE AND H. H. BARTLETT

Hevea brasiliensis, like many other tropical trees, is deciduous in just the same sense that the majority of the trees of temperate climates are. The leaves are not shed a few at a time, throughout the year, and replaced by the continuous unfolding of new ones. On the contrary, each tree sheds all of its foliage at some season of the year, and remains bare for a time, during which it is said, from analogy with the trees of cold climates, to be "wintering." The wintering period is followed by the unfolding of new leaves and the determinate growth of the twigs that bear them. Just as in an apple or a maple tree, a short period of growth is normally followed by the maturing of new wood and foliage, which remains unchanged until the next wintering season.

THE LEAF-FALL DISEASE: AN ABNORMAL ABSCISSION OF YOUNG LEAFLETS FOLLOWING LEAFING-OUT

In the early months of 1918 a severe leaf-fall disease was noted over large areas in the rubber plantations of Asahan. Within a few days after leafing-out had commenced, the new leaves, although healthy and turgid, began to fall off. In some spots almost every tree lost some of its new leaves, and many trees lost all of them. A second set of leaves was immediately produced, and frequently a third, after the falling of the second. On badly affected trees the development of new leaves from

* The work reported in this paper was done, mostly in 1918, at the rubber plantations and botanical laboratory of the Hollandsch-Amerikaansche Plantage Maatschappy, Asahan, East Coast of Sumatra, and is published with the approval of the U. S. Rubber Plantations, Inc., of New York.

dormant buds left the ends of the branches of the year as bare spurs, projecting from the young foliage, and the eventual result was a characteristic fastigiate branching.

As a result of the leaf-fall disease the trees were without functional foliage for an abnormally long period, a deprivation which in extreme cases must have seriously affected the growth of the trees and the production of latex.

When the young leaflets fall they are usually very small, rarely over one fourth of full size, and frequently less than two cm. long. The leaflets usually fall from the petiole first; the petiole itself drops from the twigs a little later. The leaflets are usually quite turgid when they fall, but are sometimes more or less withered at the tip; oftentimes they look healthy, but drop at the slightest touch. Shaking a badly diseased tree brings down a rain of leaflets and petioles. Sometimes the twigs die back for a short distance from the tip, but this is not common nor characteristic of the disease as seen in 1918.

Trees of all ages were diseased, from seedlings of one year in nursery rows up to the oldest trees on the plantations. The disease was found to be present on all types of soils, wet or dry, rich or poor. The damage was very slight, however, on good soils where the trees were in a thrifty, vigorous condition. On poorer soils, where the condition of the trees was not so good, the disease was often severe.

Fallen leaves showed no sign of fungus infection, and no lesion of any kind. In the beginning the leaf-fall was ascribed to mites by some of the Estate staff, but careful examination showed that very few trees were affected by mites. The general appearance of the fallen leaves was very different from that of leaves attacked by mites.

THE CULTURE AND IDENTIFICATION OF *Gloeosporium alborubrum*

A large series of cultures, summarized below, indicated that *Gloeosporium alborubrum* Petch was the cause of leaf-fall. In the experiments recorded below the test of identity of Gloeosporium was in every case the production of the characteristic pink spore masses. The characteristics of the mycelium, such as the habit of growth, were not relied upon. If a mycelium did not fruit, even though it appeared to be Gloeosporium, it was classed with the unidentified group.

Cultures of Gloeosporium obtained from different sources were grown on a variety of media. The medium most commonly used was made by boiling young succulent twigs and leaves of Hevea so as to make a strong decoction. To this was added a small quantity of Batak sugar,[1] which is sold in every native shop, and about three per cent of agar. This medium is easily made and supports good growths, not only of Gloeosporium, but also of other fungi parasitic on Hevea, such as Diplodia, Helminthosporium, Pestalozzia, etc. This agar is rather acid and usually one has little trouble with bacterial contamination when it is used. Gloeosporium also grows well on rice agar, bean agar, potato agar, bouillon agar, etc. It thrives on uncooked slices of manihot root, of sweet potato, of cocoanut, and of white potato. On the last of these it produced a deep, characteristic dry rot. On sterilized Hevea fruits and twigs it develops with great rapidity. No other parasitic fungus of rubber develops in culture more rapidly than Gloeosporium, or produces spores in larger number. A mycelium inoculation in an agar tube usually produces a colony which develops great masses of pink spores in from twenty-four to thirty-six hours. When a tube is inoculated with a piece of diseased tissue, spores are usually produced first on the tissue, then later on the agar. In very vigorous cultures the whole surface of the agar becomes covered with a thick layer of spores.

The mycelium of the fungus when young is pure white, but it soon darkens to grey or black especially on the surface of the agar in tube cultures. When a little older the surface tends to become very smooth and satiny.

The spores of *Gloeosporium alborubrum* are pink in mass, but

[1] Bartlett, H. H. *The Manufacture of Sugar from Arenga saccharifera in Asahan, on the East Coast of Sumatra. Ann. Report Mich. Acad. Sci.*, 21: 155–165. 1920.

seen individually under the microscope they are colorless and hyaline. They are oblong in shape, 15–20 microns long and 3–4 microns broad. At maturity they are non-septate, but become septate upon germination. The spores do not germinate in distilled water or tap water, but in weak solutions of glucose and of cane sugar they germinate within a few hours. When the spores germinate in masses, the germ tubes from the different spores fuse so as to form a single mass of mycelium. This phenomenon occurs in many fungi, but especially freely in Gloeosporium.

In our entire experience with isolations of Gloeosporium from Hevea we saw no evidence that more than one species existed in our cultures, which comprised two or three hundred primary isolations and numerous transfers. Petch,[2] however, had given three names to Gloeosporia from Hevea, two of which, *Gloeosporium Heveae* and *G. brunneum*, had been synonymized by himself. Only two names, therefore, needed to be considered in identifying our organism, these being *G. Heveae* Petch and *G. alborubrum* Petch.

Gloeosporium Heveae was described in Ceylon as the cause of leaf-fall of young nursery plants about a foot high. The leaves first turned yellow-green, then yellow, and fell off. The spores were produced in "pale brown masses" on either side of the leaf. This disease differed from the other recorded leaf diseases of Hevea in that there was a general death of the whole leaf, not of isolated patches. (As will be seen, *G. alborubrum* is capable of causing just such symptoms of infection.) In his later publications Petch states that this fungus was found only once, in 1905. It seemed strange that the very first isolation of a Gloeosporium, a fungus subsequently found to be almost omnipresent in Hevea, should have been of a form different from the one so common. It seemed so likely that the two putative species were identical that Mr. Petch's latest view of the matter

[2] Petch, T., *Die Pilze von Hevea brasiliensis* (*Para Kautschuk*). *Zeitschr. f. Pflanzenkrankheiten*, 18: 81–92. 1908.

Petch, T., *The Physiology and Diseases of Hevea brasiliensis*. London, 1911. (See pp. 171, 218, 261.)

was sought, and it was found that he, too, now doubts the validity of the distinction which he relied upon to differentiate the two species. The color of the fresh spore masses in the Gloeosporium of Hevea is characteristically pink, whence the name *G. alborubrum,* but was described by Petch as "pale brown" for *G. Heveae.* The latter form was no longer in culture and not available for comparison when Mr. Petch made his later studies of the die-back disease of Hevea twigs and named *G. alborubrum* as the causal fungus. We have used the name *G. alborubrum* in this paper as one which is certainly applicable, in order to avoid confusion should *G. Heveae,* with pale brown spores, as originally described, again come to light. It is likely, however, that the two supposed species are one, and that the use of the prior name, *G. Heveae,* would have been justifiable.

Gloeosporium alborubrum has been regarded by Petch as the cause of a die-back of Hevea twigs. His argument as to the causal relationship of the fungus to the disease seems convincing to the writers, in the light of their own experience of the numerous manifestations of Gloeosporium infection. Petch says that the die-back of green shoots due to *Gloeosporium alborubrum* has been known in Ceylon since 1905, but that no serious damage has ever resulted from the death of green shoots only, since the trees send up fresh shoots from buds lower down, and little harm is done. As he very emphatically states, however, the lesion caused by Gloeosporium affords a point of entrance to the wound parasite *Botryodiplodia Theobromae* (= *Diplodia cacaoicola* P. Henn.) which may greatly extend the die-back, or cause the death of the whole tree. (In connection with the rapid death of trees, apparently from Diplodia, La Rue[3] has shown that the possibilities of lightning injury must not be overlooked.)

[3] La Rue, Carl D., *Lightning Injury to Hevea brasiliensis. Phytopathology,* 12: 386–389. 1922.

CULTURES TO DETERMINE THE CAUSE OF LEAF-FALL

Experiment I. — On May 24, 1918, young leaves, which had lately fallen, were collected, sterilized in $HgCl_2$, 1/1000, and rinsed twice in sterile water. The base of the blade of each leaflet was then pinched off with sterile forceps and dropped into a tube of sterile agar. Forty cultures were made altogether, which gave fungi as follows:

Gloeosporium alborubrum Petch	28
Helminthosporium Heveae Petch	3
Pestolozzia Guepini Desm.	3
Diplodia cacaoicola P. Henn.	1
Fusarium sp.	1
Miscellaneous (unidentified) fungi	18

Several of the tubes gave more than one fungus. In such cases separation of the species was effected and identifications were made when possible. *Gloeosporium alborubrum* was by far the predominant type and was at once indicated as most likely to be the causal organism.

Experiment II. — Further cultures were made on June 15, 1918. Twigs from which young leaves had fallen were collected, brought to the laboratory, very carefully sterilized in $HgCl_2$, 1/100, and rinsed in sterile water. Twenty-three twigs were used and from each the tip was put into one tube, the bark from a portion of the twig some distance from the tip into another, and the wood from this portion into a third. Sixty-nine cultures were made altogether. After a week the cultures were examined and transfers of the fungi made. A number of culture tubes remained sterile even after the lapse of nearly a month. (This gave a good indication of the efficiency of the sterilization!) The sterile tubes were as follows: fifteen containing wood, four containing bark, three containing twig tips.

As might have been expected the wood of the twigs was more free from fungi than either the bark or the twig tip. The result of the cultures was as follows:

Gloeosporium alborubrum	16
Diplodia cacaoicola	4
Miscellaneous fungi	32

The miscellaneous fungi which did not fruit and so could not be identified with any certainty, comprised some eight or nine different forms, as far as their appearance and cultural characteristics indicated. Gloeosporium was again the predominating form. Some of the tubes contained more than one fungus.

Experiment III. — On June 19, 1918, cultures were made from a number of trees with leaf-fall on Boenoet Estate. A number of twigs were collected, and the petiole of one leaf from each of these twigs was also taken. These were carefully scrubbed with $HgCl_2$, 1/1000, rinsed with sterile water and put into tubes to incubate. It should be noted that the petiole used was in each case one which had actually lost leaflets from the disease. A portion of each twig was taken for culture from a point near the insertion of the cultured petiole. Twenty-five tubes were inoculated with petioles, and a like number with portions of twigs. The following fungi were produced by the cultures:

	Petioles	Twigs	Total
Gloeosporium alborubrum	21	18	39
Diplodia cacaoicola	1	2	3
Fusarium sp	0	1	1
Miscellaneous fungi of 3 types	2	3	5
Sterile tubes	1	1	2

In these cultures more than eighty per cent of the isolations were of Gloeosporium. The entire absence of bacteria and the common contaminating organisms such as Aspergillus, Rhizopus, etc., indicated that the superficial sterilization of the material was thorough. The evidence from these cultures pointed unmistakably to Gloeosporium as the cause of the disease, provided, of course, that any fungus infection could be looked upon as the cause.

Experiment IV.— On June 19, 1918, cultures were made from seedlings in the experimental nursery, the leaves of which showed a tendency to fall as soon as they developed. Twigs were cut from these seedlings and sterilized in $HgCl_2$, 1/1000, then washed and put into tubes. On the same date petioles from similar seedlings were secured from which the leaflets had fallen. These were cultured in the same way as the twigs. Where more than one fungus developed in a tube, separation of the types was made. The final result of nineteen twig and nineteen petiole cultures was as follows:

	Twig Cultures	Petiole Cultures
Gloeosporium alborubrum	3	8
Diplodia cacaoicola	0	2
Fusarium sp	3	1
Helminthosporium Heveae	3	2
Miscellaneous fungi, comprising 5 or 6 types	8	5
Sterile	1	0
Bacteria	1	1

It is noticeable that the petioles gave Gloeosporium cultures more frequently than the twigs. The petiole cultures would seem to offer a better opportunity of detecting the real cause of the disease than the twig cultures, and they indicated Gloeosporium to be the causal organism.

This completes the list of cultures made from the trees that were diseased, as manifested by the actual falling of leaves. Altogether 188 isolations were made, and of these 94 or 50% proved to be *Gloeosporium alborubrum*. The other 94 isolations showed at least 9 or 10 kinds of fungi and at least one form of bacterium.

From other work done on Diplodia, Helminthosporium, Pestolozzia, and Fusarium, it was expected that these fungi would occur frequently in the cultures made. All, except Fusarium, were known to be parasitic on Hevea, Diplodia being almost constantly present in trees of all ages; Helminthosporium and Pestolozzia being common on and in Hevea leaves. These fungi account for 15% of the isolations. The remaining 35% are divided among at least 5 or 6 unidentified fungi and bacteria. When all of these facts are considered, the evidence that *Gloeosporium alborubrum* is the cause of the disease seems almost incontrovertible, if the cause is to be found in any fungus.

In addition to the evidence given above, pertinent facts were observed in connection with an experiment in which it was attempted to secure vegetative reproduction of Hevea by rooting in moist sand cuttings made during the wintering season. The cuttings, in the most hopeful cases, put forth set after set of leaves, which quickly fell off, whereas others soon rotted, undergoing a sour fermentation. Eventually all died, and the moist portions developed masses of Gloeosporium and Diplodia spores. Some of the cuttings were cultured at the same time their leaves were falling, and Gloeosporium was isolated several times. It appeared that the cuttings made were infected when brought to the laboratory, and that under the conditions of the experiment the disease was much more harmful than on the plantation.

OTHER MANIFESTATIONS OF GLOEOSPORIUM INFECTION

Evidence was obtained that Gloeosporium is a frequent infecting organism in leaves and twigs, and that it has other manifestations than the leaf-fall of young leaves.

Leaf-curl. — On June 8, 1918, cultures were made from leaves which had reached maturity, but were badly twisted and curled. Cultures, ten in all, were made from these leaves and from the twigs on which they were borne, in the usual manner. All gave *Gloeosporium alborubrum*, without any contaminating organisms.

Leaf-spot. — On May 24, 1918, cultures were made in ten tubes from leaf-spots, which were very common on the mature leaves. The usual cultural methods were employed. Four of the ten tubes remained sterile; the other six all developed *Gloeosporium alborubrum* while two had *Pestolozzia Guepini* in addition to Gloeosporium.

Yellow-leaf. — On the same date, May 24, 1918, twelve cultures were made of leaves which had just reached maturity, but which were turning yellow in an abnormal manner. Nine of these developed Gloeosporium and the other three gave fungi which did not fruit and could not be identified.

Fruit-fall. — Not only are the immature leaves subject to abnormal abscission, but the young fruits are likewise. On May 24, 1918, ten cultures were made from fragments of fruits, which had fallen prematurely, and from the peduncles of such fruits. Four tubes remained sterile and six gave cultures of Gloeosporium.

Tip and Margin Disease. — Another sort of Gloeosporium infection on Hevea, which was not for a long time recognized as due to Gloeosporium, shows itself on mature leaves. The leaves have dead areas at the tip, and sometimes along the margins. Sometimes only the extreme tip is dead; again the whole margin may be killed, leaving the center of the leaf intact, or the infection may be so severe as to kill the whole leaf. The disease was distinct from leaf-spot, since it rarely killed regions of the leaf not connected with the leaf margin. In this type of infection one suspects complications of a physiological nature, especially since tip and margin disease seems to be most serious on bad soils. Early in 1919, almost every tree in some areas of bad soil was seen to be more or less affected. The trees most seriously attacked were wintering very early and it seems likely that the disease was responsible for this condition. In the latter part of 1919 and the early part of 1920 comparatively little tip and margin disease was seen. These observations are in accord with the fact that there was little leaf-fall from Gloeosporium after the wintering in 1919.

On January 11, 1919, diseased leaves were collected from Soengei Baleh Estate where many of the trees were affected by tip and margin disease. The leaves were thoroughly scrubbed over the entire surface with $HgCl_2$, 1/500. They were then rinsed in sterile water and pieces were cut from sound tissue just beyond the dead areas. One tube out of 39 remained sterile, but in all of the other 38 Gloeosporium colonies developed which produced spores.

The first evidence of the cause of tip and margin disease was secured in 1918 from cultures of leaves from Pamoedian Estate, near Pematang Siantar. At that time this disease was supposed to be due to Pestalozzia. This view was supported by the fact that the trees from which these leaves were taken grew along a tea field which contained many plants with "grey blight," which is due to *Pestolozzia Guepini* Desm. However, when cultures were made in the usual way, in ten tubes, all developed Gloeosporium.

On May 8, 1919, leaves with dead tips from Boenoet Estate, Asahan, were sterilized and ten cultures made from them in the manner described. On May 17, Gloeosporium had developed in all, and was producing spores. Hence from the cultural side the evidence is sufficient that this tip and margin disease, if due to any fungus at all, is due to Gloeosporium, since this fungus was isolated in 58 out of 59 cultures without any other fungus appearing. The incidence of tip and margin disease also corresponded very well with the years and the areas in which leaf-fall was very severe.

INOCULATIONS WITH GLOEOSPORIUM

The results of inoculations were not satisfactory either in 1918 or in 1919, because Hevea free from all suspicion of previous infection could not be secured. It is considered that the experiments are worth describing, however, since they show the ease with which infection is brought about and the rapid and destructive parasitism of the fungus when conditions are favorable to it. The reader should observe that we failed to reproduce the leaf-fall disease in experimentally inoculated material, but that under the conditions of high humidity which obtained in the experiment, leaf-fall was sometimes observed in the checks, which were not free from infection when brought in from the plantation. Under proper meteorological conditions the dormant or semidormant mycelium present in the twigs appears to become active, and to result in symptoms, such as leaf-fall, which are not as severe and immediately destructive as those which would follow the abundant inoculation with spores. Thus, a twig with dormant mycelium, placed in a damp chamber, developed leaf-fall, but withstood rotting and complete destruction a long time, whereas the latter result quickly followed when a similar twig under similar conditions was inoculated with spores.

Experiment I. — On April 23, 1918, twenty young shoots of Hevea with leaves just unfolding were brought into the laboratory and put into glasses with sterile water in the bottom, and covered with sterile Petri dish covers. Ten twigs were sprinkled with a suspension of Gloeosporium spores and ten were kept as checks.

On May 4, 1918, all of the ten inoculated twigs were badly rotted and all were producing spores of Gloeosporium, while all the twigs of the control lot except two were healthy. These two were infected with Gloeosporium and were producing spores. The mycelium was, of course, already present in these check twigs, which suggests that it may have been present and have developed similarly in the inoculated lot. However, the great numerical difference is strongly in favor of the conclusion that some of the positive cultures resulted from inoculation.

Experiment II. — Another set of inoculations was made on Feb. 28, 1919. Beakers of 1 liter capacity were covered with Petri dish covers. One hundred cc. of water was placed in each and all were then autoclaved. Newly developed twigs with the leaves just unfolding were cut from trees showing no Gloeosporium disease, and put into these beakers. These twigs were taken from 5 trees numbered from 1 to 5. Four cultures were made from each tree, those from tree No. 1 being designated as culture 1/1, 1/1 check, 1/2, and 1/2 check. Similarly, the cultures from tree No. 2 were culture 2/1, 2/1 check, etc. Cultures No. 1/1, 2/1, 3/1, 4/1, and 5/1 were now sprayed with a suspension of Gloeosporium spores. The checks, 1/1 check, 1/2 check, etc., were sprayed with sterile water only.

Cultures No. 1/2, 2/2, 3/2, 4/2, and 5/2 were pricked in twigs, petiole, and leaf lamina with a needle dipped in a mass of Gloeosporium spores. The corresponding checks were pricked with a sterile needle only.

On March 4, results were as follows:

A. Sprayed with a Suspension of Gloeosporium Spores

Cultures 1/1, and 2/1. Petioles and leaves rotting; stem tip rotten; petioles breaking at base, covered with Gloeosporium spores.

Culture 3/1. All leaves and young inflorescences rotten; most petioles and stem tip also rotten; covered with spores.

Culture 4/1. In worse condition than 3/1.

Culture 5/1. Very rotten; all leaves falling.

B. Sprayed only with Sterile Water

Culture 1/1 check. Strong, healthy, and turgid. Very small tufts of fungus hyphae growing from two petioles from which leaves were cut.

Culture 2/1 check. Entirely healthy.

Culture 3/1 check. Still healthy and turgid, but 3 or 4 leaves have Rhizopus growing on them.

Culture 4/1 check. Turgid. No stem or petiole rot. Inflorescences healthy; 3 leaves have Rhizopus on them, and one has Gloeosporium.

Culture 5/1 check. Stem tip rotten, covered with masses of Gloeosporium spores; one leaf with Rhizopus; remainder of specimen turgid and healthy.

C. Inoculated with Gloeosporium by Needle Pricks

Cultures 1/2 and 2/2. Petioles, leaf-blades and stem tips rotting; petioles breaking at base, covered with Gloeosporium spores.

Cultures 3/2 and 4/2. Whole tip of stem with all leaf-blades and petioles rotten.

Cultures 5/2. Stem tip rotten, covered with Gloeosporium spores; leaves covered with Rhizopus.

D. Pricked with Sterile Needle Only

Culture 1/2 check. Healthy and turgid.

Culture 2/2 check. Healthy and turgid, except that one leaf is infected with some fungus.

Cultures 3/2 check and 4/2 check. Turgid, but with several leaves covered with Rhizopus or sterile hyphae of other fungi.

Culture 5/2 check. Stem tip rotting. Some leaves with Gloeosporium and Rhizopus.

In general the checks were turgid, green, and healthy as compared with the inoculated twigs, which were brown, rotten, and flaccid. At this date it is very apparent that the inoculation has been the cause of a severe rotting which has not appeared in the controls, except in the case of tree No. 5, which appears to have had an active Gloeosporium infection at the time the cuttings were secured for the experiment.

On March 7, the greater part of the checks had developed active Gloeosporium disease. Here, however, it was much less serious than in the case of the inoculated twigs. In many cases the leaves dropped from the stem, leaving the stem green and turgid. In all the inoculated twigs the stem was severely rotted. In the checks the disease took the form (leaf-fall) which it usually has in the field. It is clear that the inoculated twigs quickly developed the destructive rot as a result of the inoculation, whereas in the checks the disease developed slowly from internal infections, which had not been able to develop in the open, due to the greater vigor and resistance of the host under field conditions.

Experiment III. — A further inoculation experiment was made by spraying with a suspension of Gloeosporium spores two seedlings which had been grown from seed in large cans at the laboratory, and were about two feet high. Two other similar seedlings were sprayed with sterile water only. In about one month the inoculated seedlings developed the tip and margin disease, to which we have already referred as one of the manifestations of Gloeosporium infection. Almost every leaf was affected on the inoculated seedlings, while on the controls no sign of the disease was found. Later, cultures were made from leaves of the inoculated seedlings and all developed Gloeosporium. This evidence was very good indeed, though the number of seedlings, all that were available, was too small.

INFECTION UNDER FIELD CONDITIONS

The precise manner in which the parasite enters the host is not known. It appears that the disease of the young leaves is not due to infection by spores, but rather to the growth of mycelium from the twig. In cases where leaves do appear to have been infected by spores, a wilting of the tip indicates the probable point of entrance of the parasite. In the greater number of cases, however, the young leaves fall while still turgid and healthy, quite without any sign of fungus injury. Here it appears most likely that a growth of mycelium from the twig causes abscission of the leaflets.

The leaf-tip and margin phase of Gloeosporium disease may follow a water-pore infection, and from leaves so infected the fungus may gain entrance to the twig, where it lives through the wintering period. After wintering, as the young leaves develop, the mycelium probably grows out into the petioles and leaves, causing abscission, if environmental conditions are suitable.

Just what conditions favor the disease is not known. The failure of the disease to develop markedly in 1919 was presumably due to different meteorological conditions from those that prevailed at the end of the "wintering" season in 1918.

The susceptibility of the trees to the disease is doubtless also determined largely by their vigor. Wherever the trees were very vigorous, leaf-fall disease, even in 1918, was hardly noticeable, but on the most unfavorable soils where the trees were least vigorous the disease was worst. It is believed that areas of thrifty rubber will not suffer much from Gloeosporium, but that in areas where the trees are noticeably lacking in vigor it may cause notable losses. As Petch has shown, twigs which are killed at the tip by Gloeosporium offer an entrance to Diplodia, which may cause further die-back. As we show in the following paper, other sequelae may be infection of wounds, decay of bark on tapping surfaces, etc. The loss of two or three sets of leaves on a tree cannot be considered negligible, because the material used in forming the leaves may seriously deplete the stored food reserve of the tree. Also, the sooner the tree is in full leaf after the wintering period, the sooner it begins to produce food for growth and yield. A delay of even a month in coming into leaf may be serious from an economic standpoint. Trees which are weakened by an attack one year, will not be in as good condition to resist the disease another year, and so declining vitality may end in loss.

PREVENTIVE MEASURES

Combative measures against Gloeosporium have not been devised. Prevention by isolation from infected areas is impossible as the fungus is very wide-spread and can grow on a large number of plants besides Hevea. Spraying would not aid, if the disease develops from within the twig, unless such spraying were continued for some years, enabling the trees to outgrow old infections. Even then, it might not be successful, and an extensive spraying campaign would probably not be embarked upon unless the prevalence of other diseases made such a procedure advisable from the standpoint of general sanitation.

It is probable that fertilizers, such as sodium nitrate, which noticeably increase the vigor of trees growing on the soils which are least satisfactory for rubber, will be of use in combating the disease, should its attacks become serious.

SUMMARY AND CONCLUSIONS

1. *Hevea brasiliensis,* in the East Sumatran plantations, and probably generally, is subject in certain seasons to an abnormal leaf-fall, characterized by the abscission of the young and apparently healthy foliage.

2. The leaf-fall disease does not occur after every wintering season. It appears to be "physiological" in the sense that the infection to which the disease is attributed produces its characteristic symptoms only when the meteorological conditions are

favorable. Even then, disease is manifest only in trees of low vigor, such as those on the poorest soils.

3. The leaf-fall is attributed to *Gloeosporium alborubrum* Petch, which is probably the same as *G. Heveae* Petch. (In the latter suggestion Petch concurs with the writers.) *G. alborubrum* was originally described by Petch as the causal organism of a die-back of young Hevea twigs in Ceylon, and *G. Heveae* Petch as the cause of yellowing of the leaves of seedling Hevea. About two hundred isolations in Asahan show only one species of Gloeosporium on Hevea, which conforms to the description of *G. alborubrum,* infects both leaves and twigs, and is associated with several abnormal conditions of the host, such as leaf-curl, a leaf-spot, yellow-leaf, premature fall of young fruits, etc.

4. Under usual plantation conditions Gloeosporium has been shown to exist in a dormant condition in many Hevea trees. The leaf-fall and the fall of young fruits appear to be caused by growth of the mycelium from the older tissues into the new parts.

5. Primary infection, from spores, appears to cause the condition which we have termed "tip-and-margin" disease. This type of disease may be experimentally produced by spraying Hevea with a suspension of spores. From such primary infections the fungus doubtless enters the twigs from the leaves.

6. Those symptoms of Gloeosporium infection, particularly leaf-fall, which we attribute to the perennial mycelium in the host, rather than to primary infection of the leaves by spores, have not been produced by inoculation because of the lack of experimental material which was certainly free from the disease.

7. The inoculation of Hevea cuttings, in damp chambers, with Gloeosporium spores, produces severe rotting of all young tissues, — stem, fruit, inflorescence axes, and leaves. The same results are secured in uninoculated cuttings when placed in a damp chamber, but after a much longer incubation period, provided that the cuttings contain dormant Gloeosporium mycelium.

8. *Gloeosporium alborubrum* grows readily and sporulates abundantly on a variety of artificial media, and grows on living

tissues not only of Hevea but also of manihot (roots), sweet potato (roots), cocoanut (endosperm), and white potato (tuber). On the latter it produces a deep and characteristic dry-rot.

9. None of the manifestations of Gloeosporium infection in Hevea are alarming except in trees of low vitality. The complex of diseases associated with this fungus will be best combated by any measures of cultivation and fertilization which maintain the trees in a high state of vigor.

UNIVERSITY OF MICHIGAN

DIPLODIA DISEASE OF *HEVEA BRASILIENSIS*

CARL D. LA RUE AND H. H. BARTLETT*

SEED DETERIORATION

Diplodia cacaoicola P. Henn. is well known to be one of the ubiquitous fungi of tropical plantations, infesting a great variety of hosts. In *Hevea brasiliensis* it produces a die-back of the twigs and branches. Our own first experience with Diplodia on the United States Rubber Plantations (Asahan, East Coast of Sumatra) was in connection with the rapid destruction of Hevea seeds by fungi. Apparently sound seeds retained their germinating power only a short time after dehiscence of the capsules, and within a month or so the contents had completely rotted.

For the first cultures a nutrient agar was used, which contained a decoction of green Hevea bark as a nutrient medium. The seeds were soaked in alcohol, and the alcohol on the surface ignited, after which they were cracked with sterile pincers, and portions of the kernels put into the agar tubes. In a few days several of the tubes showed a growth of a fluffy white mycelium which grew with great vigor, and eventually proved to be *Diplodia cacaoicola* P. Henn.

To determine whether or not the fungus was pathogenic in nature, and so capable of causing sterility of the seeds, inoculations were made on living fruits and roots of several cultivated plants. The results are shown in Table I. All fruits, and roots

* Most of the work reported in this paper was done jointly by the authors in 1918, at the rubber plantations and botanical laboratory of the Hollandsch-Amerikaansche Plantage Maatschappij, Asahan, East Coast of Sumatra, but was extended in 1919 by the first author. It is published with the approval of the U. S. Rubber Plantations, Inc., of New York.

were sterilized externally with $HgCl_2$. They were then halved with a sterile knife, and put into sterilized tumblers and covered with halves of Petri dishes similarly sterilized. One half was inoculated and the other kept without inoculation as a control.

TABLE I

INOCULATIONS WITH DIPLODIA MYCELIUM ON LIVING TISSUES

Substratum	Number of inoculations	Number of checks	Results	Checks
Hevea fruits	6	3	Fruits rotten and covered with mycelium	Checks much later were self-inoculated from small black spots and began to rot
Banana.....	2	Half of each as check	Rotten and covered with long hanging mycelium	1 sterile, 1 beginning to show Penicillium
Egg plant...	1	Half as check	Completely rotten	Sterile and healthy
Potato......	2	Half of each as check	Odorless wet rot, completely decomposing cells, but leaving starch grains intact	Sterile and healthy
Sweet potato	2	Half of each as check	Very slight rot	Sterile
Manihot....	4	Half of each as check	Completely rotten; wet rot. Odor yeasty and fruity. Mycelium long, hanging, shaggy	Sterile

The results given in Table I were, with the exception of the egg plant and the sweet potato, apparent in three days, although complete rotting did not take place until later. The surprisingly rapid development of the organism as a parasite showed that it was capable of causing serious injury, and we ascribed to it the rapid deterioration of Hevea seeds. It was some time, however, before any of the cultures produced spores enabling the fungus to be determined. In the meantime, however, a large group of isolations accumulated, from many parts of the Hevea tree, which produced the same characteristic wet rot of the potato (tubers grown in the Karo highlands) and showed the same vegetative characters. These cultures eventually fruited, and proved to belong to *Diplodia cacaoicola* P. Henn.

CULTURAL CHARACTERISTICS OF DIPLODIA FROM HEVEA

Growth of Diplodia mycelium on the white potato is extremely rapid, and a very thorough wet rot is caused so that the whole content of the potato may be poured out from the skin. All of the cells are broken down; even the cell walls are disintegrated, and only the starch grains are left intact. This is in sharp contrast with the behavior of the organism on sweet potato on which it is scarcely parasitic at all, a curious result in view of the fact that at least one species of Diplodia causes a serious rot of sweet potatoes. Potato cultures came to be used regularly as a test for Diplodia, since the Diplodia rot could be easily distinguished from that of other common organisms. *Gloeosporium alborubrum*, for example, caused a deep dry-rot.

The organism was cultivated on various media in the attempt to make it fruit so that its identity might be discovered. On rice agar, French-bean agar and potato agar, it grew with great rapidity. On rice agar it developed a stronger growth of aerial mycelium than on either of the other two media; potato agar gave the smallest growth. In all cases the mycelium was at first entirely white, but soon it began to darken and finally all of the mycelium became entirely black. This blackening took place soonest on rice agar and was longest delayed on bean agar.

After some months, the first fruiting on agar finally occurred. Cultures had previously been made on sterilized wood of Hevea, and also on sterilized green twigs. The last two culture media were used specially to induce fruiting, since Diplodia fruits abundantly on dead Hevea wood under plantation conditions. In tubes, however, no fruiting took place for some weeks, even on wood. Sterile water had been put into the tubes of both wood and twigs to prevent their drying out too much. Only after this water had evaporated were spores produced. Apparently a drying out of the medium was necessary to induce fruiting. This was also apparent in the agar cultures, since they never fruited until they were old and the agar was rather dry. At the beginning of our work transfers to fresh tubes were made too frequently, and favorable conditions for sporulation did not occur.

Several strains were cultured on sterile Hevea latex, by the transfer of mycelium from agar cultures. On latex the mycelium never showed the very vigorous growth characteristic of the fungus when growing on a rich substratum. At the end of a month the latex was not coagulated nor discolored. The rubber globules in all the tubes had risen to the top like cream, but had not coagulated, and could be readily distributed again by shaking. It was found that this latex could be boiled without causing it to coagulate. The experiments showed that latex was not a suitable culture medium for Diplodia; they are mentioned because they indicated that infection of Hevea bark by Diplodia without other accompanying organisms would probably not affect latex flow. As will be seen, Diplodia was found as a frequent parasite on the wounded bark of the tapping cut.

BARK INFECTION BY DIPLODIA

Miscellaneous isolations led to the conclusion that certain diseased conditions of the tapping area of the bark might be due to Diplodia infection. In order to determine the reaction of the tree to the fungus, a series of saplings were inoculated as follows:

Series A: Inoculations of unripened stems. — The green stems of twenty-two saplings between 90 and 120 cm. high were split about fifteen centimeters below the apex and a tuft of fungus mycelium was inserted. The two halves of the stem were then allowed to come into contact again, thus enclosing the mycelium in the interior; twenty-two other trees were split in the same way, but these were kept as controls, no mycelium being inserted.

Series B: Inoculations of ripened stems. — Ten saplings with a diameter of about five centimeters were inoculated near the ground level. The bark was sterilized as far as could be done with alcohol. Then a vertical cut was made with a tapping knife in such a way as to leave a flap of bark attached. The cut extended to the cambium throughout most of its length, reaching into the wood in some places. In each cut a tuft of mycelium was inserted and the flap of bark tied back into place so as to cover the mycelium.

Ten other trees of similar size were kept as controls. These were treated exactly as the others except that no mycelium was inserted in the cuts.

Five days after the smaller seedlings, *Series A*, were inoculated, an examination was made. The top of one was entirely withered and falling over. All the other inoculated seedlings were diseased, but not so badly. All the controls were free from disease and were healing nicely.

The ten larger saplings, *Series B*, were not examined until twenty-four days after inoculation. At this time they showed little disease of cortex, but all exhibited serious cambium and wood infection, the cankers being about 7.5 cm. long, 1 cm. wide, and 0.5 cm. deep. Streaks were to be seen in the wood which indicated that the infection had traveled a much greater distance in the wood both up and down the stem. The edges of the inoculated wounds were not healing.

The checks showed no disease and the wounds on the stem were nearly closed by growth from the sides. When examined the wood showed, in a few cases, a slight discoloration where the cut had been made, but elsewhere was pure white and quite without any of the streaks indicative of diseased vessels.

Series C. — A mature Hevea tree was now inoculated in two different places. The bark of the tree appeared to be perfectly healthy before the inoculation was made. Two control cuts were also made, but not inoculated.

After twenty-four hours no external change was noticeable. During the second twenty-four hours an exudation of yellow transparent latex was seen. Forty-eight hours after inoculation one inoculated cut was opened and explored. A well defined case of bark rot was found and the infection had spread for a distance of 10 cm. up and 10 cm. down the stem. The typical black streaks in the wood before described extended somewhat farther than this. The inner bark was very soft and mushy, but the outer bark was unchanged except near the cut. The control cuts were perfectly healthy.

Four days after inoculation the other cut was explored. The cambium was black and dead for a considerable distance and the black streaks in the wood were very extensive. The control cuts were still perfectly healthy.

RECOVERY OF DIPLODIA FROM THE EXPERIMENTALLY INFECTED TISSUES

Cultures were made from infected regions of a number of the saplings and the mature tree. All of these cultures gave the original fungus while controls did not. Determinations of identity at this time were made by testing all the strains on a series of living substrata, of which potato proved the most useful. Up to this stage of the work none of the strains had fruited, and so large a series of cultures had accumulated that particular attention had to be directed to inducing sporulation, in order to make final identification of the cultures and eliminate many of the strains. Among the cultures were strains supposed to be Diplodia, because they had been isolated from Hevea twigs showing die-back, from Hevea wood which subsequently produced Diplodia spores abundantly on standing, and from decaying Manihot root. These strains were all identical in vegetative characters, and agreed with the authentic Diplodia grown on agar tubes from spores from dead Hevea wood and Manihot roots. Although the evidence was strong that all of the strains were Diplodia, it was necessary to bring all to fruiting. As already indicated, this problem solved itself for it was found that fruiting took place in neglected cultures which had exhausted the culture medium and had dried out. All of the cultures fruited when handled in accord with this discovery, whether inoculations had been made on agar, or on Hevea twigs.

After it had been ascertained definitely that the strains were all Diplodia, new inoculations were made with strains from various sources, to ascertain if their effect on living Hevea would be the same. Two year saplings in the nursery were used. The bark of each tree was washed with alcohol and then more alcohol was poured on the bark and ignited. The bark so sterilized was now excised with a sterile knife, and shallow cuts over an area of three to four square centimeters were made. A sterile cotton pad moistened with water was laid against the wound with an inoculum of mycelium growing on agar between the cotton and the wound. Sterile cloth was now wrapped around the tree to hold the cotton in place, and the whole bandage was sealed with melted paraffine. Twenty-one trees were inoculated and nine were cut and bandaged without inocula as controls. Several strains of the fungus were used.

Ten days later an examination of the inoculated trees and controls was made. All of the controls showed a bright green color on the wounded surfaces and were healing rapidly. The inoculated wounds, on the contrary, all looked black and unhealthy. The bandages were not replaced, but the wounds were left open to the air and the light.

A month from the time the inoculations were made the wounds were explored. The severity of the infections varied somewhat as might occur if the strains differed in virulence, as was doubtless the case. Although the wounds originally made did not extend to the cambium, in almost all cases the bark was killed to the cambium. In one case the death of the bark did not affect the cambium extensively, but injury to the wood was evident. In another case the bark was killed only in spots. In general, the bark was completely killed under the whole area of the original wound, and a thick pad of rubber had formed between the bark and the wood. The injury to the bark was not observed to extend into the unwounded bark for a distance greater than 2 cm. The wood under the wound was in most cases dead and discolored. The characteristic brown and black streaks in the wood extended down the trees for a distance of 11 to 25 cm. and up the trees for 13 to 51 cm.

The control trees showed a healthy growth of bark over the wound. A thin shell of dead bark was formed outside, and under this was a new green layer. Both bark and wood were perfectly healthy and no discoloration was to be seen in either.

The pathogenic organism was re-isolated from wood and bark of the inoculated trees, but from none of the check trees, tissues of which were cultured in the same way that the infected ones were. Precautions were of course taken to culture wood and bark that were not injured when the inoculations were made.

DIPLODIA AS THE CAUSE OF 'MOLDY ROT' OF THE TAPPING SURFACE

Our chief work with Diplodia was directed toward determining its relation to the condition known as 'moldy bark.' This condition was first noticed in rubber divisions where coolies had been trained to tap, and where large areas of bark had been tapped within a few days, offering great opportunity for infection by wind-borne spores or by spores from rain water running down the trunk. 'Moldy bark' was not confined to the training divisions, however, but was likely to occur, especially during the wet weather, wherever the tapping was too deep, or whenever the tapping was so unskillfully done that the cambium and wood were exposed.

The study of 'moldy rot' is complicated by the occurrence of many fungi, including some that are merely saprophytes, on the diseased and dead bark. Once the bark is dead, a number of fungi develop on and in it within a very short time. We have examined specimens of 'moldy rot' which had spores of all the following fungi on the surface in sufficient quantity to indicate that all were fruiting vigorously: Diplodia, Fusarium (three forms), *Gloeosporium alborubrum*, *Helminthosporium Heveae*, *Penicillium sp.* Conidia of four or five unidentified types were also found.

It is quite possible that 'moldy rot' may often start with the killing of wounded bark by sun and wind; the infection with fungi follows. The wood underlying an area affected with 'moldy rot' is, however, usually full of brown or black streaks characteristic of Diplodia disease. Commonly these streaks can be traced to an old wound near the diseased area. As long as the bark above this wood remains unharmed the fungus appears not to be able to invade the bark. When the bark is tapped, however, the fungus is enabled to produce a severe infection on the tapped surface and often causes the death of the bark over extensive areas. It has already been shown that these black streaks may extend in the wood for a very considerable distance above and below the original infection; thus an area may be infected and die from 'moldy rot' although at least a half meter above or below the old wound where the fungus originally gained entrance. In such cases, naturally, the depth of tapping and the general condition of the tree are important factors in determining susceptibility.

'MOLDY ROT' EXPERIMENTALLY CAUSED BY INOCULATION OF TAPPING CUTS BY DIPLODIA

Fourteen mature trees on Soengei Boenoet Estate were used to test the supposed connection between Diplodia and 'moldy rot.' In one place on each tree the bark was opened by a tapping cut extending over about one third of the circumference of the tree. The bark was then tapped away for a vertical distance of about seven centimeters. A similar area was now opened on the opposite side of the tree, but either nearer the ground or farther from it than the first cut, to prevent infection spreading from one cut to the other. The second area was to serve as a control. It was possible to have the control on the same tree because experience had shown that infection by Diplodia spread very slowly in the bark, and that its spread was vertical rather than lateral in either wood or bark.

Inoculations were made with Diplodia spores shaken up with water and applied with a brush to one cut on each tree. Both inoculations and controls were left uncovered, so that controls were exposed to chance infection.

Ten days after the inoculations were made the trees were examined. All the inoculated cuts were infected, but on two of the trees the infections were in the surface layers only. On all the others the infections were very severe and large areas of bark were killed.

Of the control cuts, seven were infected, but one only lightly. In no case were the infections so severe as on the inoculated cuts. It was, of course, to be expected that a certain number of these large fresh wound surfaces would be infected by spores from the plantation. It is, therefore, not surprising that several of the checks were infected in the same way as inoculated trees, but less seriously. The experiment was a crude one, but showed definitely that there was no doubt about the deleterious results of infection with Diplodia spores. Seven of the control cuts remained perfectly healthy, and on the other seven the areas of dead bark were scattered and not confluent, as in the case of the inoculated areas.

The infections did not extend far in the bark beyond the limits of the surface exposed by the cuts. In the wood, however, the spread was somewhat greater. The wood was completely killed under the inoculations for a depth of from 2 to 3 cm. and for an area considerably greater than that inoculated. The brown and black streaks in the wood could be detected for a distance of from 30 to 70 cm. above and below the inoculations.

After ten months had elapsed the inoculations and controls were again examined. All the wounds which had been severely infected were still unhealed, and the dead wood was exposed. The areas where the bark had died were slowly being covered by bark growth from the sides. The bark around the wounds for a distance of 4 or 5 cm. was discolored and diseased. The wood was found to be dead over an area somewhat greater than the original wound. In some cases the wood had been killed for a depth of from 5 to 10 mm. and for a distance of 25 or 30 cm. above and below the wound. This dead wood had been covered over, except under the original wound, by about 1 cm. of new wood formed by the cambium. This new wood appeared entirely healthy, and the diseased or dead wood could be seen only by cutting away the sound new tissue. The characteristic brown and black streaks in the wood, caused by the vessels being filled with the black mycelium of the fungus, could be readily traced for a distance of 50 cm. below the wound and about a meter above the wound.

In those of the control wounds which were uninfected at the time of the first examination, no trace of disease could be seen in the bark or under the bark in the wood. The wood under the wound and beyond it was pure white and free from the characteristic brown and black streaks.

Table II shows the condition of the inoculations and controls after one year. (One of the fourteen trees had died and had been removed.)

TABLE II

RESULTS OF DIPLODIA INOCULATIONS AFTER ONE YEAR
(*Inoculated area and control on same tree*)

No. of tree	Inoculated cut	Control cut
1	Unhealed area 2.5 × 7.5 cm.	Unhealed area 2.5 × 2.5 cm.
2	Unhealed area 7.5 × 10 cm.	All healed
3	Unhealed area 10 × 20 cm. (No healing)	Unhealed area 6 × 7.5 cm.
4	All healed	All healed
5	Unhealed area 2.5 × 12.5 cm.	All healed
6	Unhealed area 1.5 × 2.5 cm.	All healed
7	3 unhealed wounds each 1 × 1 cm.	All healed
8	Unhealed area .75 × 4 cm.	All healed
9	Unhealed area 1 × 4 cm.	2 unhealed wounds 1.2 × 2.5 cm.
10	All healed	Unhealed area 2.5 × 2.5 cm.
11	Unhealed area 2 × 4 cm.	Unhealed area 2.5 × 2.5 cm.
12	Unhealed area 1.5 × 9 cm.	Unhealed area 1.2 × 8.7 cm.
13	Unhealed area 1.5 × 9 cm.	All healed

The area of unhealed wounds on the inoculated cuts was about 565 sq. cm. as compared with 75 sq. cm. on the control cuts. This comparison gives an idea of the area of unhealed wound surface on the inoculated cuts that may have been due to sun scald and accidental infection under the particular conditions existing at the time of the experiment. The difference is so great that no doubt remains regarding the possibility of injurious Diplodia infection on the tapping surface. When it is considered that the die-back of young branches from Diplodia is very common; and that the dead branches afford an abundant source of spores, which may be carried down the trunk and into the tapping cut by rain, or disseminated by wind, it is clear that some phase of Diplodia disease should be expected on the tapping cut. This phase appears to be what has been called 'moldy bark.'

DIPLODIA INFECTION AND 'BLACK THREAD'

Frequently a Diplodia infection extends down or up the tree from a wound as a narrow line in the wood. Where the bark over this infection is tapped deeply the infection comes out and causes the death of the bark along this line. Such infections are commonly called 'black thread,' and are said to be due to *Phytopthora Faberi*. Dozens of specimens from such infections were cultured, but no Phytopthora was ever found, although Diplodia was always present. Many of the more extensive infections on the tapping surface which are attributed to 'black thread' disease, are apparently due to Diplodia infection. It would be rash indeed, in view of the large amount of work which has been done with the disease known as 'black thread,' to assert that it is merely a phase of Diplodia infection having nothing to do with Phytopthora. In Asahan, however, the experimental induction by inoculation with Diplodia, of a disease like that usually called 'black thread' and the failure to culture Phytopthora from cases diagnosed as 'black thread,' lead us to suspect that even if a 'black thread' caused by Phytopthora does exist, other fungi also cause a similar condition. One of them is Diplodia. We venture to state that 'black thread' is often one of the sequelae of 'moldy rot,' and that both, as they occur in Asahan, are to be attributed to Diplodia.

FINAL INOCULATIONS TO PROVE RELATIONSHIP OF 'MOLDY ROT' AND 'BLACK THREAD' TO DIPLODIA

As a final test of Diplodia as a cause of 'moldy rot' and 'black thread' an experiment was made on one hundred trees at Soengei Baleh Estate. On each tree a tapping cut was opened over one third of the circumference and about 4 cm. of bark was cut away. The trees, numbered from 1 to 50, were now inoculated by brushing the tapped surface with a suspension of Diplodia spores in water, so that every part of the exposed surface was covered. The trees numbered from 51 to 100 were similarly brushed with water alone. All the cuts were left exposed as in the previous experiment.

Eight days after the inoculations were made the trees were all examined. All of the inoculated trees showed some infection, which was generally serious. Of the controls nearly half showed no infection whatever, only one showed a bad infection, four showed a moderately bad case, and the remainder showed only slight infections, which in most cases did not extend to the cambium. Eight of the infections were traceable to old 'black thread' on the tapping surface.

Table III gives the results of the inoculations.

TABLE III

RESULTS OF FIFTY INOCULATIONS OF THE TAPPING CUT WITH DIPLODIA SPORES

Character of infection	Inoculated trees	Control trees
Very bad infections	22	1
Moderately bad infections	17	11
Slight infections	11	24
No infection	0	21
Traceable to old 'black thread' infection	1	8

The results of the infections were evaluated; very bad infections were scored as 3, moderately bad infections as 2, and slight infections as 1. When the inoculations were scored on this basis they gave a total of 111 out of a possible 150 while the controls scored only 35. The average degree of infection would be 70 % for the inoculations and 20 % for the controls.

An examination of the infections and controls was made ten months after the inoculations to see what were the after-effects of such infections. The data are given in Table IV. Aside from the conditions recorded in the table, it is to be recorded that bad infections were always accompanied by black streaks running up the wood, — the 'black thread' symptom.

TABLE IV

CONDITION OF TAPPING CUTS TEN MONTHS AFTER INOCULATION WITH DIPLODIA

	Inoculated trees	Control trees
Total number of open wounds	27	13
Total area of open wounds	54 sq. inches	25 sq. inches
Total number of trees with excrescences	41	19
Total number of trees with bad excrescences	27	5
Total number of trees entirely healed and without excrescences	3	23

While a certain number of trees, both among those inoculated and the controls, probably became infected by exposure to spores carried by the air, and while it is also likely that some wounds in each class were caused by the action of the sun on the exposed cambium, the results show clearly that the inoculation itself was the cause of very great damage to the trees.

It should be noted that this experiment at Soengei Baleh was made on trees in rather poor condition and hence the results are probably worse than they would have been on very vigorous trees.

An examination of the inoculated and the control trees on S. Baleh showed that almost all of the trees which were seriously infected by inoculation with Diplodia spores, when tapped below the inoculated area developed 'moldy rot' and lost large areas of bark from this cause. The uninfected trees were very little subject to this disease and practically no bark on these trees died from this cause.

A further test of the relation of Diplodia to 'moldy rot' was made on the trees on Boenoet Estate which had been used in the experiment already described. The original inoculations had been made nearly a year before and healthy wood had covered over that filled with the dark streaks. Five of these trees which had severe infections after inoculations were chosen for the

TABLE V

RESULT OF TAPPING AREAS OF BARK ABOVE INFECTED AND UNINFECTED WOUNDS

No. of tree	Above infected wound	Above uninfected wound
1	Bark over nearly all tapped area dead. Wood dead to depth of 1 mm. Typical Diplodia infection	Bark healthy No Diplodia
2	All bark of tapped area dead. Wood below tapped bark dead to depth of 2 mm. Typical Diplodia infection	Bark healthy No Diplodia
3	All bark of tapped area dead. Wood under dead bark dead to a depth of 2 mm. Wound spreading upward in bark	Small area of bark dead, but no Diplodia infection. Possibly tapped too deeply
4	Bark healthy	Bark healthy
5	All tapped area infected and most of bark on this area dead. Wood below dead to depth of 2 mm.	Bark infected with Diplodia but not dead. Wood sound under tapped area. Probably due to outside infection

experiment. An area of bark about 30 cm. above the wound was now tapped away both on the inoculated side and on the control side of the tree. It was expected that if 'moldy rot' is caused by the fungus coming out from the wood, it would develop on the tapped areas above the old infected wounds and not on those above the uninfected wounds. It was feared that the thickness of healthy wood which had developed outside the infected wood would vitiate the experiment by preventing the infection of the bark. Table V shows the results of the experiment.

It appears that the fungus is able to penetrate the healthy layer of wood and cause the infection of the tapped bark overlying healed-in 'black thread.'

SUMMARY

Cultures of the same Diplodia (generally called *D. cacaoicola* P. Hennings) which causes die-back of Hevea have been obtained from practically all parts of the Hevea tree.

What appears to be the same fungus is the cause of a storage rot of the Manihot tubers. It is actively parasitic on many other plants, notably on the potato, in which it causes a rapid and destructive wet-rot of the tubers.

Diplodia is believed to be an important cause of seed deterioration in Hevea.

The condition known as 'moldy rot' of the tapping area may be experimentally produced with Diplodia spores.

Diplodia infection in the wood brings about black streaks in the wood, due to the fact that the mycelium, which is black, follows the vessels upward and downward from the point of inoculation. These black streaks are not distinguished, even by experts, from a disease called 'black thread' which is usually attributed to *Phytopthora Faberi* Maubl.

Tapping of uninfected bark above infected wood is followed by an active infection ('moldy rot') of the bark.

It is fair to conclude that much of what is called 'black

thread' disease is due to Diplodia, and that the fungus primarily concerned in moldy bark is likewise Diplodia.

Diplodia causes little permanent damage in healthy trees, since the infections heal over and are eventually deeply buried. The several infections which it causes are most serious in trees growing under unfavorable conditions, or subjected to unusually extensive wounding.

UNIVERSITY OF MICHIGAN

STOMATAL BEHAVIOR OF PLANTS IN THE GREENHOUSE IN WINTER

HUGH B. SMITH

INTRODUCTION

History. — As is well known, the first critical study on the opening and closing of stomata was made by von Mohl (10) in 1856. Following his work was that of Schwendener (11) who determined the mechanism of stomatal opening. The theory of stomatal mechanism, as recognized by von Mohl in 1856 but worked out by Schwendener in 1881, is the accepted one at the present time. Other well-known workers in this field are Kohl (4) in 1886, Leitgeb (6) in 1886, and Francis Darwin (1) in 1889. In more recent times, the particularly prominent workers are Lloyd (7) in 1908, Laidlow and Knight (5) in 1916, Gray and Peirce (3) in 1919, and Loftfield (9) in 1921.

The general conclusion that can be drawn from all of their work is that light governs the opening and closing of stomata if conditions are optimum for moisture and temperature. However, when conditions become less favorable for the usual alternative movements, the behavior becomes less regular, as shown by Leitgeb (6) and Loftfield (9). Moreover, when conditions become extreme, the stomata may be closed all day and open all night, as shown by Loftfield (9) for *Medicago sativa*. The unfavorable conditions causing this unusual behavior were drought and high temperature. Loftfield discovered also that the stomata of *Triticum vulgare*, *Zea mays*, and *Pennisetum glaucum* are generally open in the forenoon and closed in the afternoon in the summer, the afternoon closure being due to a decrease in leaf-cell turgor accompanied by evidences of wilting. Gray and Peirce (3), however, found in the case of *Secale cereale* and *Avena fatua* in the dry climate of California that there was a closure in the afternoon without evidences of decrease in the leaf-cell turgor. This is the normal behavior of these plants in that region and may be a varietal adaptation for the conservation of moisture.

The Problem. — Most of the data so far published on the opening and closing of stomata have been worked out for plants grown in the open in the summer. Dr. F. C. Newcombe suggested that I determine the stomatal behavior of plants grown in the greenhouse in the winter. Throughout the winter months the light for the latitude of Ann Arbor, Michigan, is meager, but it is reduced still more in the greenhouse where the plants I studied were growing. This greenhouse is on the University of Michigan campus, and, although placed on the south side of one of the buildings, other buildings near by cut out about an hour of sunlight at the close of day.

METHOD OF OBSERVATION

There are a number of methods for determining the degree of stomatal opening. These are the cobalt chloride method of Stahl (12), the horn hygrometer method and porometer method of Francis Darwin (1), the porometer method as improved by Laidlow and Knight (5), the method of stripping off of epidermis used by Lloyd (7), and the method of direct observation suggested by Lloyd, but first extensively used by Gray and Peirce (3). The last named method is one in which the stomata are observed on leaves which, without being removed from the plant, are held on the stage of the microscope. The leaf is illuminated from the mirror below with sunlight or with artificial light. Gray and Peirce (3) believed that a cooling cell, such as a Soyka flask filled with water and wired beneath the stage, is necessary on hot days.

I used this method of direct observation with some modifications. I believe that artificial light is better than sunlight even though the sunlight is strong. This is the case because, with the use of artificial light, one can obtain a strong light below the stage and have, at the same time, but little light above the stage.

I am sure that a cooling cell is not necessary for observations made in the winter.

DETERMINATION OF OPEN AND CLOSED CONDITION

After deciding on the method of observation, it became necessary to determine how a stoma should look when it was closed as opposed to open. I decided that a stoma was closed if I could see no light slit in the dark area between the guard cells, and that it was open if a slit of light did appear in the middle of the otherwise dark area. It should be remarked that even in a single field of view of the leaf, that is, over an area of less than a square millimeter, some of the stomata will frequently be open and some of them will be closed. This is particularly true for observations made at a time not far removed from the natural time of opening and closing. Moreover, this is more likely to be the case on cloudy days than on clear days. This variation of stomatal behavior has been reported by all of the workers on the subject. For determining the open and closed state I used the standard of Lloyd (7), that is, I eliminated the extremes and considered the condition of the majority from two or three fields of view on a leaf. Moreover, I was careful to select a leaf sufficiently isolated on the plant so as not to have been shaded before the time of observation.

OBSERVATIONS

Behavior on Dark Days. — The data presented in Table I show the behavior of stomata on a dark day. The data obtained on February 19 were selected for the table because the results on that day were particularly striking. The time given in the following table and throughout the paper is mean time. The species are as named in L. H. Bailey's *Standard Cyclopedia of Horticulture.*

TABLE I

SHOWING BEHAVIOR OF STOMATA ON DARK DAY

(O = Open, C = Closed)

Feb. 19, 1922. Foggy A.M. Dark and rainy P. M.

SPECIES EXAMINED	9:00 to 9:30 A.M.	10:00 to 10:30 A.M.	2:15 to 2:45 P.M.	4:15 to 4:45 P.M.
Tradescantia fluminensis Vell.	O	O	O	O
Nephrolepis exaltata Schott	O	O	O	O
Adiantum cuneatum Langs. & Fisch.	C	C	C	C
Primula kewensis Wats.	O	O	C	C
Fuchsia speciosa Hort.	C	C	C	C
Phaseolus vulgaris L.	C	C	C	C
Beta vulgaris L.	C	C	C	C
Coleus Blumei Benth.	O	O	O	C
Pteris longifolia L.	C	C	C	C
Antirrhinum Orontium L.	C	C	C	C
Asparagus asparagoides Wight.	C	C	C	C
Rosa sp.	O	O	C	C
Begonia lucerna Hort.	O	O	O	C
Pelargonium Endlicherianum Fenzl.	C	C	C	C
Impatiens Sultani Hook.	C	C	C	C

The data in Table I show that the stomata of *Tradescantia fluminensis* and *Nephrolepis exaltata* were open all day; those of *Adiantum cuneatum, Fuchsia speciosa, Phaseolus vulgaris, Beta vulgaris, Pteris longifolia, Antirrhinum Orontium, Asparagus asparagoides, Pelargonium Endlicherianum* and *Impatiens Sultani* were closed all day; those of *Primula kewensis* and *Rosa sp.* were open during the forenoon, but closed during the afternoon; and those of *Coleus Blumei* and *Begonia lucerna* were open till the latter part of the afternoon. Thus it is shown that the stomata of some species may remain closed all day, that other species may have open stomata even on dark days, and that still other species behave as intermediates between these two extremes.

Specific Differences in Behavior on Dark Days. — Data obtained on the behavior of stomata for the five dark days occurring during a period of twenty days, from January 23, 1922, till February 12, 1922, prove the same variation in behavior of species as indicated in Table I. On February 1, the time of sunrise was 7:15 and of sunset 5:11. Seven species were followed during the twenty days. The seven species studied are *Tradescantia fluminensis, Nephrolepis exaltata, Primula kewensis, Adiantum cuneatum, Coleus Blumei, Fuchsia speciosa,* and *Phaseolus vulgaris.* At about 4:15 P.M. on February 1, the first dark day, the stomata of *Tradescantia fluminensis, Nephrolepis exaltata, Primula kewensis,* and *Coleus Blumei* were open while those of *Adiantum cuneatum, Fuchsia speciosa,* and *Phaseolus vulgaris* were closed. This same relationship existed also at the time of the other observations, namely, February 2 at 4:45 P.M., February 7 at 4:15 P.M., February 9 at 11:30 A.M., and February 10 at 10:20 A.M. and 4:50 P.M.

Behavior on Clear Days. — In order to show results obtained under conditions of greater light intensity, observations made on clear days are presented. Data obtained on February 16, which was a clear day, show, for the fifteen species listed in Table I, that the stomata of all the species were open when first examined at about 9 to 9:30 A.M. This is in striking contrast to results shown in Table I for the dark day of February 19, when, at the same time of day, eight species had closed stomata, as has already been pointed out.

Specific Differences in Behavior on Clear Days. — Data obtained on the opening and closing of stomata on clear days during the period from January 23 till February 12 show that the stomata of *Tradescantia fluminensis, Nephrolepis exaltata, Adiantum cuneatum, Fuchsia speciosa, Coleus Blumei, Beta vulgaris, Primula kewensis,* and *Phaseolus vulgaris* open earlier and close later on clear days than they do on dark days; they show further the same variation in behavior of species as is shown in Table I. For example, on January 25 at 7:10 A.M., ten minutes before sunrise with clear sky, the stomata of *Tradescantia fluminensis* and *Nephrolepis exaltata* were open while those of *Adiantum cuneatum*, *Beta vulgaris*, *Coleus Blumei*, *Fuchsia speciosa*, *Primula kewensis*, and *Phaseolus vulgaris* were closed. This same relationship held true also on February 8 at 5:15 P.M. Data on these and other clear days — observations on ten clear days in all — show that the stomata of *Tradescantia fluminensis* and *Nephrolepis exaltata* open about sunrise and close shortly after sunset, while the stomata of the other six species just named open later and close earlier than this.

Behavior on Partly Cloudy Day. — The data presented in Table II show the behavior of stomata on a day which was cloudy part of the day and clear part of the day.

TABLE II

SHOWING BEHAVIOR ON DARK DAY FOLLOWED BY CLEARING SKY

Feb. 3. 1922. Snowy and cloudy till 3:00 P.M. Clear after 3:00 P.M. for rest of day

SPECIES EXAMINED	7:35 to 7:50 A.M.	10:05 to 10:20 A.M.	12:25 to 12:40 P.M.	3:10 to 3:25 P.M.
Tradescantia fluminensis Vell.	O	O	O	O
Nephrolepis exaltata Schott.	C	O	O	O
Adiantum cuneatum Langs. & Fisch.	C	C	C	O
Fuchsia speciosa Hort.	C	C	C	O
Phaseolus vulgaris L.	C	C	C	O
Beta vulgaris L.	C	C	C	O
Coleus Blumei Benth.	C	C	C	O

Observations for Table II were made on February 3 which was dark until 3:00 P.M., but clear after that time for the rest of the day. The data show for *Tradescantia fluminensis* the same behavior as was shown in Table I, that is, the stomata were open all day; they show that the stomata of *Nephrolepis exaltata* opened some time between 7:50 A.M. and 10:20 A.M. and, when once open, remained open for the rest of the day; for *Coleus Blumei*, that its stomata opened some time between

10:20 A.M. and 12:40 P.M. For *Adiantum cuneatum, Fuchsia speciosa, Beta vulgaris,* and *Phaseolus vulgaris* the data for February 3 indicate the same variation in species as the data for February 19, and the data for the two days show further that the stomata of some species, which may be closed throughout a very dark day, may, on the other hand, open late in the afternoon if the sun should come out at that time.

Behavior at Night. — Results by Leitgeb (6) and Loftfield (9) showing night opening and day closure of stomata on plants growing under conditions of low moisture and high temperature would lead one to suspect that stomata might, in conditions other than low moisture and high temperature, be open during the night if they were closed during the day. I made observations at different times of the night ranging from about 6:30 P.M. on some days until as late as 1:30 A.M. on other days, and not once did I find the stomata open after dark.

CONCLUSIONS

A. From a study of the eight species, namely, *Tradescantia fluminensis, Nephrolepis exaltata, Adiantum cuneatum, Fuchsia speciosa, Coleus Blumei, Primula kewensis, Beta vulgaris* and *Phaseolus vulgaris*, the following conclusions are drawn:

1. The stomata of *Tradescantia fluminensis* and *Nephrolepis exaltata* open on clear days in winter about sunrise and close shortly after sunset; moreover, they open earlier and close later, hence require less light, than those of *Adiantum cuneatum, Beta vulgaris, Coleus Blumei, Fuchsia speciosa, Phaseolus vulgaris,* and *Primula kewensis.*

2. The stomata of *Adiantum cuneatum, Fuchsia speciosa, Beta vulgaris,* and *Phaseolus vulgaris,* which are the ones in the list above having closed stomata on cloudy days, may open their stomata late in the afternoon if the light intensity increases at that time.

B. From a study of the fifteen species, namely, *Tradescantia fluminensis, Nephrolepis exaltata, Adiantum cuneatum, Primula kewensis, Fuchsia speciosa, Phaseolus vulgaris, Beta vulgaris,*

Coleus Blumei, *Pteris longifolia*, *Antirrhinum Orontium*, *Asparagus asparagoides*, *Rosa sp.*, *Begonia lucerna*, *Pelargonium Endlicherianum* and *Impatiens Sultani*, the following conclusions are drawn:

1. As far as my observations went, stomata are never open at night in the winter.

2. The stomata of *Tradescantia fluminensis* and *Nephrolepis exaltata* are open all day on cloudy days, even when the day is very dark, but they open later and close earlier on cloudy days than on clear days.

3. The stomata of *Coleus Blumei*, *Begonia lucerna*, *Primula kewensis*, and *Rosa sp.* require more light than the two foregoing species for the opening of their stomata, and usually open them during only the lighter part of a dark day.

4. The stomata of nine of the fifteen species listed remain closed all day on particularly dark days. These nine species are *Adiantum cuneatum*, *Fuchsia speciosa*, *Phaseolus vulgaris*, *Beta vulgaris*, *Pteris longifolia*, *Antirrhinum Orontium*, *Asparagus asparagoides*, *Pelargonium Endlicherianum* and *Impatiens Sultani*.

C. The general conclusion drawn from this work is that light is the governing factor in winter for the opening and closing of stomata, but the intensity necessary for opening differs for the various species.

I wish to acknowledge my gratitude and indebtedness to Dr. F. C. Newcombe under whose guidance this study was made.

University of Michigan

BIBLIOGRAPHY

1. Darwin, Francis. 1898. Observations on Stomata. Phil. Trans. Roy. Soc. London B., 190: 531–621.

2. Darwin, Francis, and Pertz, D. F. M. 1911. New Methods of Estimating the Aperture of Stomata. Proc. Roy. Soc. London B., 84: 149.

3. Gray, J. and Peirce, G. J. 1919. The Influence of Light upon the Action of Stomata and its Relation to the Transpiration of Certain Grains. Am. Jour. Bot., 6: 131.

4. KOHL, F. C. 1895. Zur Mechanik der Spaltöffnungsbewegungen. Acad. Caes. Leop. Car. Abstract Bot. Jahresb., 23: 414. 1895.

5. LAIDLOW, C. J. P., and KNIGHT, R. C. 1916. A Description of a Recording Porometer and a Note on Stomatal Behavior during Wilting Ann. Bot., 30: 47.

6. LEITGEB H. 1886. Beiträge zur Physiologie der Spaltöffnungsapparate. Mitth. Bot. Inst. Graz., 1: 123–184. Abstract Bot. Jahresb., 14: 61. 1886.

7. LLOYD, F. E. 1908. The Physiology of Stomata. Carnegie Inst. Wash. Pub., No. 82.

8. LLOYD, F. E. 1913. Leaf Water and Stomatal Movement in Gossypium and a Method of Direct Visual Observation of Stomata *in situ*. Bull. Torr. Bot. Club, 40: 1.

9. LOFTFIELD, J. V. G. 1921. The Behavior of Stomata. Carnegie Inst., Wash., Pub. No. 314.

10. MOHL, H. VON. 1856. Welche Ursachen bewirken die Erweiterung und Verengung der Spaltöffnungen? Bot. Zeit., 14: 697.

11. SCHWENDENER, S. 1881. Ueber Bau und Mechanik der Spaltöffnungen. Monatsber. Kön. Akad. Wiss., Berlin, 27: 833–867.

12. STAHL, E. 1894. Einige Versuche über Transpiration und Assimilation. Bot. Zeit., 52: 117.

FOREIGN INVESTMENT AND FOREIGN TRADE

C. E. GRIFFIN

I

American business men are today vitally concerned with the relation of foreign trade to foreign investment. On the one hand we are constantly told that extension of long-time credits to our impoverished customers is the prime essential for a revival of our lagging export trade, and on the other hand the fear is often expressed that when we come to collect the principle and interest on these advances and those already made, we shall be confronted with an excess of imports. Thus these questions which have in the past primarily concerned academic students are now the subject of active discussion by more broad-visioned business men.

In the present discussion four questions will be raised: What is the normal relation of trade and investment? Through what steps have the great creditor countries passed in their evolution toward their creditor status? What is our present position in this normal line of development? Is our present position in the world of borrowers and lenders likely to be temporary or permanent?

Nearly everyone will agree that there is a close relation between foreign investment and foreign trade. As to the precise nature of that relationship, however, there is a diversity of opinion. The relation may be direct and immediate or it may be indirect and less obvious. When an American manufacturer extends long-time credit to make possible a purchase of his goods, the relation is, of course, direct. Perhaps a Chinese railway company desires locomotives, rails and other capital goods. It issues bonds and presents the bonds to an American manufacturer in direct payment for the goods. As the manufacturer has immediate need of funds for the conduct of his business, he sells these bonds through investment institutions to individual investors in this country. In this way American investors have placed their funds abroad and by so doing have clearly advanced American foreign trade.

In another case an American bank may underwrite an issue of foreign securities, making it one of the conditions of this financing that the equipment or supplies purchased with these funds shall be bought from specified manufacturers with which the bank has an intercorporate relationship. In such a case, which probably is not as common as is popularly assumed, there is again a clear relationship of investment to trade.

Again, although no formal restrictions upon the expenditure of funds may be made, the investors may see to it that engineers and operating officers are selected from those of their own acquaintance and nationality. This requirement would probably not be made from any patriotic bias, but merely because of the greater confidence which investors have in projects controlled by their own associates. But English engineers will prefer English types of equipment, and in the production of these types of equipment English manufacturers will have an advantage and will normally get the orders.

There have been instances of government regulation of investments with a view to encouraging the expenditure of funds in the home country. Thus the French government maintains a control over the listing of securities on the Paris Bourse and has at times made such listing contingent upon an agreement by the borrower to spend at least a part of the funds in France.

All of these factors whereby investments aid trade are, it will be noted, of a more or less artificial nature. They are mentioned here primarily to indicate that, although they may at times be effective, they do not provide the fundamental explanation of this close relationship between the two. Fundamentally, the close relation between investments and trade rests upon that elementary principle that international payments in the long run must be paid with goods or services. This is true of capital investments as well as of indemnities and other forms of international payments. Therefore, it must follow that the loan of a given sum of money can be accomplished only by exporting goods or services in return for a promise of repayment at a future date or by refraining from accepting imports of goods or services which would otherwise have come to the investing country. The export of goods may follow directly from the loan of capital, as in the instances mentioned above, or it may take a more indirect course.

Let us assume that American investors lend fifty million dollars to Argentine railroad and industrial companies. These funds may conceivably be spent entirely in England. If the normal relation of exports and imports is to be maintained in England (and there are in fact very weighty factors working for the continuance of such a balance of trade), this will normally induce Englishmen to purchase an equivalent amount of goods abroad. England may spend the equivalent of these funds in Italy, which country in turn may make purchases elsewhere. But, unless the original loan were effected by an actual shipment of gold, the American balance with the rest of the world will only be stabilized when the United States has parted with goods of a value equivalent to the loan. The original purchasing power of fifty million dollars may be passed from hand to hand, but ultimately it must be utilized in the purchase of American goods.

This automatic relationship between investments and exports works itself out through the rates of international exchange. The original loan is very likely made by authorizing the Argentinians to draw upon us for the amount represented by the loan. In other words the supply of dollar exchange in Buenos Aires is increased without a corresponding increase in the demand for American dollars. As a result the price of dollars declines in terms of pesos. This ordinarily encourages Argentine merchants to purchase in the United States and discourages American importers from buying in Argentina. This continues until American exports have exceeded American imports by the amount of the loan, when the exchange rate again comes back to normal. Through the operation of arbitrage transactions this lowering of the price of American dollars is likely to be diffused among all the trading countries of the world. It thus becomes easier for foreigners generally to purchase American goods and correspondingly more difficult for Americans to purchase foreign goods. Thus, either through a rise of our exports or a decline of our imports or through both of these, our exports come to exceed our imports by approximately fifty million dollars.

Another mode of expressing the same idea is to say that the loan of funds creates a supply of dollar exchange. Now dollar exchange can be used in the last analysis only in payments to the United States. No matter through how many hands it passes, it still remains American purchasing power, and hence if it is to be utilized at all, it must finally come back to the country of its origin.

Very often it is impossible to discover the effect in specific exports, but the causal relationship is none the less real, for we do know that over a period of time there is a very accurate correspondence between the placing of loans and other items on the debit side of the international trade balance and merchandise exports and other items on the credit side. During the early period of our national existence we borrowed large sums from England, and the material aspect of this borrowing was the excess of imports over exports, though there were no requirements made that the money should be spent in England. This relation between borrowing and imports is, of course, the converse of the relationship which we are now discussing.

It is fortunate that the wide development of international commerce makes possible an indirect rather than a direct relation between loans and exports. Otherwise we could only loan capital to those countries which happened to desire the specific goods and services which we could offer, and which we could offer on more advantageous terms than could other exporting countries. As it is, we can lend capital funds to Argentina, although the goods which we can most advantageously export and which foreigners can most advantageously purchase here may be wheat for the people of Europe and raw cotton for the mills of England. Argentina, if it thinks best, can spend the borrowed funds in

Germany or England while we can furnish to Germany and England the goods required to offset the loan. The only requisite to such lending is that there shall be a balance between loans to all countries and our exports to the world at large.

Inasmuch as the goods which will be exported under these conditions of unrestricted purchases will presumably be those lower in price here than elsewhere, and inasmuch as these are the goods in which our labor is most effectively employed, it follows that our profits from the sale of these goods will be greater than they would be in the sale of goods in which the comparative effectiveness of our labor was less, and it also follows that the amount of capital which we can loan on this basis is larger than it could be if it were arbitrarily required that the money borrowed here be spent here. It certainly is advantageous to the borrower to use his funds freely for the purchase of goods and services in the cheapest markets. Aside from any question of abstract justice and the rights of the borrower to use his funds as he sees fit, it is difficult to see how we can, in the long run, be benefited by requiring our needy borrowers to use their funds in another and less economical manner.

Many American manufacturers and some government officials have recently been inclined to criticise our bankers for making foreign loans without restricting their employment to the purchase of American goods. If the preceding statement of the relation of investments to exports be correct, it would appear that the tying of strings to our investments in this manner is not only unnecessary, but would be likely to check the development of overseas investment and limit its usefulness in fostering our export trade. The requirements made by bankers and other investors should, in most cases, be limited to such matters as the rate of interest, the security for the loan, and, perhaps, to the purpose of the loan, whether it is to be used unproductively, as in wars, or productively, as in the development of railroads and industries. The lenders should not primarily concern themselves with the direction of the expenditure of the funds. The fundamental maxim of trade, "Buy in the cheapest market and sell in the dearest market," is the basis of wise economic policy here as elsewhere.

II

There is a logical line of development through which on *a priori* grounds we would expect countries to pass as they became industrialized. A country just starting in its industrial development will have need of capital. The opportunity for profitable investment will establish an interest rate high enough to attract funds from older countries. This implies an excess of imports into the new country. This borrowing process will continue until an approximate equality in the earnings of capital is established between the borrowing and lending countries. The excess of imports, however, will not continue as long as this. Inasmuch as interest must be paid upon the total amount invested up to any given date, and the borrowings as they affect trade include only the investments during a given year, there must come a time when interest payable in any given period will exceed the new capital invested in that period.

This point marks the beginning of the second or interest-paying period. During this period exports will exceed imports by ever increasing amounts. If the borrowed funds have been used in wars or in other non-producing ways, the country will not advance out of this interest-paying stage. Exports will exceed imports to pay interest and the standard of living of its people will become progressively lower. But let us assume that the borrowed capital goes to build up the transportation lines, and helps to develop the producing capacities of the country. Two results will follow: (1) The volume of goods produced will become much larger, and thus the country is enabled to meet its interest payments by exporting a part of its produce without an immediate return of merchandise or services; (2) Because of the newly developed capacity of the country to produce and save, it is enabled to supply part of its own capital, and the interest rate begins to decline.

When this decline in the interest rate proceeds far enough, there will be a tendency to buy back the home securities. This new tendency marks a third stage and brings with it a change in the balance of trade, exports now exceeding imports. This buying back of the home securities is likely to precede the purchase of foreign securities because of the preference which investors usually show for the securities of their own country. This stage, like the second, is marked by a favorable merchandise balance. It ends when the home securities held abroad are balanced and no more than balanced by foreign securities held within the country. The continued accumulation of capital, however, will force a lower return upon capital, lower interest levels than prevail in other less developed countries.

How long it will take for a country to reach this condition of depressed interest rates depends, of course, upon the supply of and demand for capital within the country itself. Thus a country with large area and large natural resources can find profitable employment for its accumulating capital within its own borders for a long time, while a small country will quickly reach a saturation point. A marked advance in the state of the arts, such a change as that from hand power to steam, or from steam to electricity, may create such a demand for new capital to put the new methods into effect that interest rates will, for a time, be maintained in the industrial country. But these new methods of production also make it possible to bring into existence an ever increasing supply of capital, so that the only condition which could prevent an ultimate decline of interest within the industrial country relative to the rates abroad would be constant and rapid progress in the uses of capital. As soon as this rapid progress abates (as it sooner or later must abate), there will be a tendency for capital to flow from the industrial countries to the relatively undeveloped parts of the world.

This brings us to the investing period which, like the preceding period, will be characterized by a favorable balance of trade. This period is likely to be brief, for in a short time interest flowing back will exceed the annual output flow of investments. Thus we are brought to the fifth and final stage in which the earnings of capital invested abroad will exceed the new capital investments during any given period of time. This new relation of capital items will be reflected in an excess of imports over exports.

Our hypothetical country has thus passed through a complete cycle beginning as a new undeveloped country borrowing capital from the older countries, and as a result having an "unfavorable" balance of trade, through the interest-paying period, through the period of cancellation of the foreign debt, through the period in which it in turn starts to lend abroad, which periods are marked by a "favorable" balance and on to the interest-receiving stage in which again it returns to an "unfavorable" balance. This is the last and final stage, for, while all the others set in motion forces which will ultimately lead to another stage, this one can continue indefinitely. By describing this as the interest-receiving stage we do not mean to imply that further investment ceases. In fact the annual investments abroad may become larger every year and exports of merchandise may very well reach higher levels than in any preceding stage. But the characteristic of this period is an excess of interest receivable over new investments and of merchandise imports over exports.

To these generalizations, based so far upon purely *a priori* grounds, certain qualifications must be made. First, the rate at which a country will evolve through these stages cannot be predicted. It will depend upon the supply of its natural resources, the initiative and intelligence of its people and their habits of thrift, its institutions, — political and social, — its freedom from wars and other disturbances and no doubt many other factors. Second, a country may be set back or advanced in this line of evolution by war or some other abnormal disturbance. Thus a country in the final stage may be set back to the first stage. On the other hand, a country which is in the interest-paying stage may, in such a time and in order to sell its goods, be advanced quickly to the capital-exporting stage and perhaps to the final interest-receiving stage. If, however, the national equipment and the industrial and commercial organization of the nations remain essentially undisturbed, such changes are likely to be only temporary and we may look for a return to the relative positions existing before the disturbance.

III

A line of development similar in its broad aspects to that described above is discernible in the history of the chief industrial countries. England started upon its development toward its modern position at a time when available capital was more plentiful at Amsterdam and certain other continental centers than in England itself. During the seventeenth century large sums, according to the standards of those days, were loaned by the Dutch for internal improvements in England. From the rather scattered data of that period it appears that from the middle of the sixteenth to the end of the seventeenth century, England had an unfavorable balance of trade such as would have been expected in a developing country borrowing capital from foreigners. In spite of the rapid accumulation of wealth during the eighteenth century this foreign borrowing proceeded. Adam Smith estimated that in 1762 some £18,000,000 of foreign capital was invested in England. Mr. C. K. Hobson shows that at that time the interest rate in Holland, which was the source of most of the borrowed funds, was commonly much lower than in England.

During the eighteenth century, however, interest payments on the previously invested capital apparently exceeded the new loans, for during this century, in spite of the fact that England continued to borrow, her imports exceeded her exports. This excess of exports progressively increased, especially during the Napoleonic wars, and enabled England, first, to cancel its foreign debt and, secondly, to lend its own capital abroad. In many ways the experience of England growing out of these disturbances on the continent was analagous to our own experience in the recent war. After the Napoleonic wars England's favorable balance continued, and her foreign investments correspondingly increased. By 1840 the interest flowing back from these accumulated investments and the earnings of British shipping combined to create inflowing payments in excess of annual investments. On the trade side this resulted in an excess of imports over exports, which has continued to the present time.

In the case of France and Germany, we do not have a complete cycle because these countries have not, at least in modern times, drawn upon outside countries for capital for their industrial development. While the capital investment in both countries has been greatly augmented in the past 200 years, most of it has been produced at home. English capital was invested in railroads in both France and Germany, but these investments were largely offset by loans from these countries to foreign governments. As a result we have only the latter phases of our cycle represented in the economic history of these countries. France had a favorable balance of trade from 1716, the earliest date for which data are available, to 1860, with the single exception of the period of the Napoleonic wars, when its balance was unfavorable. From 1861 to about 1870 exports and imports were almost equal, while from 1871 to the present, France has been in the final or interest-receiving stage with imports exceeding exports.

Germany had from the beginning of the last century up to 1880 a favorable balance of trade with the exception of the years about 1872 while the French indemnity was being paid. Beginning about 1890 and continuing to the present time, Germany's imports have exceeded her exports. This excess unfavorable balance is explained by inflowing interest in excess of loans, although in more recent years shipping earnings have helped to swell this import balance. It should be noted in the case of each of these European countries that new capital investments have been growing progressively, but in none of them have these new investments been large enough to offset the interest flowing back.

The history of our own balance of trade well illustrates the relation between capital transactions and trade. Our capital borrowing period was long, extending from 1822, the first year for which sufficient data are available, to 1873 with two interruptions. The first interruption of foreign investment here was occasioned by the panic of 1837 in which many European investors lost heavily. Confidence in American securities was not renewed until about 1849, when the buying of our securities again began. It was interrupted for a time during the Civil War only to be renewed with increasing volume and perhaps even recklessness after the war. The borrowed capital of this period went largely into railroads, many of which were almost entirely financed abroad. It is almost certain that this borrowing proceeded at a more rapid rate after 1873 than before, but by that time interest payments came to exceed the new investments and our "unfavorable" balance turned to an excess of exports over imports.

It will be recalled that in our discussion of the cycle of investments we found three successive stages which are marked by a favorable merchandise balance: first, the stage of interest payments; second, that of buying back securities; and third, that of lending abroad.

The whole period, 1873 to 1914, was marked principally by interest payments, that is to say, we remained in the first of these three phases. There was a short period in the early nineties when American securities were repurchased from European holders. This, however, was due to the fact that foreign investors were frightened by our industrial depression and by our free-money movement, and American securities soon flowed back to Europe after the darkest part of our depression was past. Apparently, the American supply of capital was not yet great enough to make it profitable to dispense with European aid. Again, after 1899, American securities, especially railroad stocks and bonds, were repurchased in large volume and some Americans fancied that we had reached a stage of capital saturation comparable to that of the European countries. But after 1907 European capital again flowed into the country, so that, according to an estimate by Professor W. Z. Ripley, the indebtedness of the United States to Europe in 1914 was greater than in 1899. All this means that by 1914 we had not truly entered upon the repayment of our debt and certainly not upon the phase of foreign investing.

The outbreak of the war in 1914 led to a rush of European investors to dispose of their American securities on the New York market. This movement assumed such proportions that it became necessary to close the New York Stock Exchange and rates of sterling exchange advanced to unprecedented levels. This repurchase of securities by American investors continued and foreign loans were made so that by 1917 the balance of indebtedness was approximately even. This cancelling of our indebtedness was made possible by a great increase in our favorable balance of trade. In the three pre-war years our annual average favorable balance was about 500 million dollars, which was offset by interest payments, expenses of our tourists abroad and other items. In the fiscal year 1915 it was one billion, in 1916 two billion and in 1917 three and one-half billion dollars. About 1917 we entered upon the investment stage proper. Since then our large favorable balances have been attended by foreign loans until to-day our net investment abroad is probably about twelve or thirteen billion dollars.

IV

The question suggested by this progress is, "Are we likely to continue to invest abroad and thus enter in the near future into the final interest-receiving period with a permanent excess of imports over exports?" The most popular answer to this question seems to be an affirmative one. It is implied in the frequent assertion that New York has surpassed London as the financial center of the world, and that the American dollar is to supersede the pound sterling as the currency of international finance, that we must look forward to a period of unfavorable balances. Admitting that we have made a remarkable record in lending capital abroad, we find certain considerations which suggest that a continuation along this line is by no means assured.

In the first place, such a sudden change in our credit position as we have experienced in the last few years has occurred before, only to be followed in a very few years by a return to the old status. Our experience in the first few years of the Civil War is a case in point. At the opening of that war large amounts of European capital, principally British, were invested in this country in the Northern States. At the first news of secession English investors were frightened. The view was quite commonly held in England that the federal government could not successfully put down the rebellion, and no doubt the open sympathy of the well-to-do classes of England for the cotton-growing South also played a part. At any rate, American securities were dumped on our market. These were rapidly taken up at low prices. David A. Wells, writing in 1869 as special commissioner of the revenue, asserted that by 1863 the United States exhibited a clean national ledger in respect to international indebtedness. Note how quickly the change had come about and then note a further statement by the same writer to the effect that by 1869 foreign capital had flowed back to this country at such a rate that we were indebted to Europe to the amount of about one and one-half billion dollars.

Another example of the brief existence of an essentially abnormal tendency is found in the repurchase of American securities during the depression of the nineties. Writers of the time freely predicted that we had passed from our debtor position and would henceforth rival England as an investing country. But in the industrial boom following 1900 the demand for capital at home was so great that foreign capital was invested here in ever increasing amounts. These instances are mentioned to suggest that the mere fact that a marked change in our creditor position has come about almost over night is no evidence that we are to continue along these new lines. In fact, the very suddenness of the change and the extraordinary conditions surrounding it give weight to the presumption that it is only temporary.

Secondly, the fact that by far the largest part of our loans to foreigners have been made by the government should make us wary of attaching much importance to our new position. A recent estimate by Mr. T. W. Lamont indicates that the total amount of European government loans subscribed by the American public has amounted to only two and one-half billion dollars and of this amount only about $800,000,000 is now outstanding. If we compare this with the eleven billion dollars owing to the United States government, it will be clear that our large foreign investments have come into existence through government loans which were made for political and military reasons. The funds involved in these loans were raised by the sale of Liberty bonds, and we all recognize that in the sale of these bonds the appeal to the investors was not primarily economic. That is to say, we have raised the money and made the foreign loans, not because of the higher interest rates abroad than at home, but rather for patriotic reasons, stimulated by effective propaganda. A permanent creditor position must rest upon essentially economic factors, i.e., upon the relative profitableness of investment at home and abroad.

Thirdly, a large part of our huge war-time balance of trade has arisen from abnormal advances in price. Our ability to produce in excess of our consumption is not indicated by the values of exports over imports in the past few years. If it did indicate this one might well expect that the United States would become permanently the world's leading creditor country. During the war period, in which our exports naturally exceeded our imports, prices rose rapidly. As a result the excess of the value of exports over imports was inflated. Further, the prices of exports advanced more rapidly than the prices of imports, and this also augmented our favorable balance.

Lastly, the condition necessary for a capital-exporting country may be briefly noted. The fundamental factors are, first, the present supply of, and demand for, capital, and, second, our capacity for capital production on the one hand and the potential opportunities for its use on the other hand, — both of these factors to be taken in connection with the effective demand for capital abroad.

With these factors in mind, compare the conditions in England with those in the United States. England has a population of some 45,000,000. The workers are, for the most part, employed in manufacturing and are aided by modern equipment. They are thrifty, and the distribution of income within the country is favorable to capital accumulation. Further, this beehive of industry has been working, producing, saving and developing the country for centuries. Now if we remember that the land area available for this development is about equal to that of the State of Michigan, it is not surprising that investments at home have proceeded so far that the earnings of capital are low, nor that fully one half of new capital accumulation flows abroad in search of less developed fields where the marginal return to capital will be greater.

But the conditions here in the United States are essentially different. Our capacity for capital production is great, considerably greater than is England's, no doubt, but the demand for capital is immeasurably greater. The future demand for capital at home can be estimated by the natural resources available for exploitation, the number of the population and its probable increase in relation to the land area and natural resources, and the natural aptitude of the people for industrial employments. Judging by European standards of intensive development of a country, we have in this country merely scratched the surface. We have a population still small in relation to our land area, but one which is growing and whose natural rate of growth can at any time be supplemented by lowering the restrictions upon immigration. The American people have that mental quality which is apparently the most important in modern industry — a genius for organization and for directing armies of workers. The opportunities for profitable investment of capital here are surely great, and it will be long before any large part of our new capital will have to seek abroad for higher interest rates.

There was a time in the nineties when interest rates in this country were low and many thought our capital must find profitable investment abroad. But since that time has come a development which has well been called a new industrial revolution. The application of electrical power to industry and the development of new forms of business organization have been the immediate causes, but the industrial growth resulting from these factors has been more marked in the United States than elsewhere because of the wide stretches of undeveloped or comparatively undeveloped land and resources. If the signs of the times in the trend of population to the cities and the changes in the nature of our exports and imports are to be believed, this is destined to be an industrial country. Further development along these lines (and such development seems inevitable) will call for large amounts of capital and will defer for a long time the coming of a condition of affairs in which our interest rate will be sufficiently below the world rate to induce a very large and sustained export of capital.

But, it may be said, large parts of the world are still undeveloped. These regions are constantly demanding new capital. Who is to supply it if not the United States? I think the answer can safely be made that new capital for foreign investment will come from Europe, as in the past. England particularly is likely to overcome its present difficulties rather soon. Before the war England had foreign investments of some twenty billions of dollars. Recent estimates indicate that only about one quarter of the total foreign investment was sold during the war. It must be remembered that while England's American securities found a market in New York, a large part of her investment was in India, China, South America, and other countries which did not have the requisite funds for repurchasing securities, and another large part of the foreign investment was in the belligerent countries themselves or in their colonies. Hence, partly by necessity, a large part of this foreign investment still remains in British hands.

True, immense debts have been incurred, but these are largely internal and hence do not affect the standing of the kingdom in relation to the outside world, while the large external indebtedness, to the United States for example, is largely offset by credits extended during the war to the other belligerents. As a net result, England is still a creditor, with her present standing somewhat impaired, but with her capital-producing power largely intact. And that capacity for producing and exporting capital was unequalled by any other country and was rapidly increasing. England's export of capital, that is, the annual net increase in its foreign capital investment, advanced rather steadily from £11,000,000 in 1902 to about £200,000,000 in 1912. It is conservative to say that by 1914 one half of England's new capital was flowing abroad. In view of these facts it is not unreasonable to expect that England will, before many years have lapsed, reassume her position as the world's greatest lending country and will largely supply the backward countries with needed capital.

Certainly there is little in our national policies and popular point of view that marks us as a great creditor country. We still view foreign investment as a temporary makeshift to enable our embarrassed customers to buy from us. We are constantly concerned with the repayment of these investments. A creditor country does not expect to collect its principal; it is only concerned with the annual interest, and expects to reinvest its capital as old loans mature. Our high tariff policy is certainly not a characteristic of a creditor.

In brief, it appears that by a set of entirely fortuitous circumstances we have suddenly become a great lender, in some respects we might say against our own wishes. But it may be doubted if we are ready for such a position. For the present, we have owing to us annual payments of interest estimated at some $700,000,000 to $800,000,000 a year. These may result for a time in an unfavorable balance of trade, although even this unfavorable balance may not be as large as is often assumed, in view of the large invisible items of American tourists' expenses, remittances of recent immigrants and the like. Although an unfavorable balance of trade may appear when and if interest payments on the large government loans begin, it may well be doubted if we shall increase our investments. Rather we may expect that sooner or later the old relative positions of Europe and the New World will reappear and that we will, from year to year, absorb almost as much foreign capital as we lend abroad.

University of Michigan

THE REVENUE ACT OF 1921

F. E. ROSS

As it would be impossible to consider carefully all the aspects of the new revenue act within the limits of a reasonably short paper, I have decided to take up only three or four of the more important points.

Two very important sections of the new act are those known as the net loss provision and the section relating to the taxation of income received through the sale of capital assets.

With one exception, it has been the rule in the past to consider the limits of a twelve-month period decisive in the determination of income. Each year's situation was considered by itself as having no connection with any other period, and while from an accounting standpoint this view was entirely rational, justice at times demanded that a longer period should be established as the true test of whether or not income had been received. For example, one corporation after sustaining heavy losses for two consecutive years, during the third year either because of luck or better management earned a large profit. The profits of the prosperous year were not sufficient to wipe out the losses of the two previous years, but were large enough to carry the computation of the company's tax into the excess-profits brackets, and the resentment of the owners at not being permitted to take the previous year's losses into consideration can readily be imagined.

In the new act the net loss provision has the effect of providing a taxable period of three years instead of one. This effect is realized, not by permitting returns to be made only once in three years, but by allowing the taxpayer to treat a net loss incurred during any taxable year after December 31, 1920, as a deduction from the net income of the succeeding year and if necessary to apply the balance against the income of the second succeeding year.

Of this provision the various interests have only the most outspoken praise to offer and so far as one can see there can be no objection raised to it.

The new provisions relating to the taxation of capital gains are an attempt to remedy a situation which, it has been claimed, has prevented to a large extent the selling and trading of property, and which, if this is so, could rightly be denounced as interfering with industrial activity. In the past, income from the sale of property was treated the same as all other income, and in the case of corporations might be taxed as high as 50 %, or, in the case of individuals, as high as 73 %. Without doubt rates as high as these would cause one to think twice before entering into a sale or trade when a large precentage of the profit would have to be turned over to the government. Profits of this sort are of an essentially different nature from business profits in that their actual or constructive receipt can be delayed without much risk, whereas ordinary industrial profits must be taken when the opportunity offers or be forever foregone. It follows that unusual business profits, which are largely fortuitous, can be taxed relatively high without greatly affecting the decisions of entrepreneurs, but equally high taxation of the proceeds of capital conversion would unquestionably discourage transactions of this nature.

The present law sets $12\frac{1}{2}$ % as a limit for the taxation of income from the sale of property owned for more than two years and which is not used by the taxpayer for personal purposes, for instance, as a dwelling. As a limitation, this provision is at present of interest only to individuals, as the maximum tax for corporations is now also $12\frac{1}{2}$ %. It would be most interesting to know if possible just how much effect this provision will have in stimulating selling and trading of the kinds of property concerned, but statistics on this point will doubtless be very difficult to obtain.

The most important and far-reaching changes are the repeal of the excess-profits tax and the increasing of the corporation income tax from 10 % to $12\frac{1}{2}$ %. The increase of the income tax rate is of course necessary to supply in part at least the revenue which heretofore has been raised by the special tax on corporations whose rate of profits was above 8 %.

These changes mark the complete abandonment of progressive taxation in so far as corporations are concerned. The wisdom of such action is questionable, lightening as it does the tax burden of some corporations and increasing it for others. Now that federal expenditures and hence the need for large revenues have been substantially curtailed, it is but fitting that the federal levies should be lessened.

On individual incomes relief has been granted which affects practically every taxpayer from the married man, with an income of $2,500 to $5,000 whose tax has been reduced by twenty dollars plus a further reduction of eight dollars for each dependent under eighteen years of age, to the recipient of a million dollar income whose surtax is now $470,960, a reduction of $112,550.

But the burden on corporations has been rearranged in a most peculiar manner. Congress, faced at once by the need of revenue and by numerous demands for the abolition of the excess-profits tax, tried to gain favor by repealing the much despised tax and at the same time make up the resultant loss by increasing the corporation income tax rate. The result is most startling. Take for example two corporations each with a income of $20,000, corporation A with an invested capital of $100,000, and B, $220,000. In 1921 corporation A would have paid in all $3,600, and B, $1,800. In 1922 each would pay $2,250, a reduction of 37 % for A, but an increase of 25 % for B. In other words corporation taxes have been reduced for those most able to pay and increased for those least able.

The excess-profits rates were without doubt too high in view of the decreased need for revenue. Especially was this felt to be true in cases where business expansion demanded that the profits be reinvested in the enterprise, and hence the accumulation of large sums for the payment of taxes was felt to be unduly burdensome.

But although a considerable degree of simplification has been secured by eliminating invested capital as one of the tax-determining factors, it would seem that greater justice would have been attained by reducing the excess tax to 10 % and leav- the income tax at 10 % as before. In the preceding illustration such action would have reduced A's tax by 25 % and left B's unchanged.

There may be some question as to whether a ratio between corporate income and invested capital is a proper basis for taxation or whether it is a test of ability to pay. The fact remains, however, that it provides the best method that has been devised so far.

In the case of corporations, size of income cannot alone determine ability to pay unless we define ability to mean simply the power to pay and have something left, without regard to justice, which should be considered. In one case a corporate income of $100,000 may only be sufficient to pay the investors 1 % or 2 % or even less, while another corporation with an income of $50,000 may be returning its owners 50 %. Justice and the present-day trend of public opinion would seem to demand that a corporation enjoying the second situation should pay more proportionally than the other.

Not only has no regard been given to this aspect in the new law, but it has been practically denied. In the past all corporations were granted a $2,000 credit against net income in computing the income tax. This credit is now refused to corporations whose net income is over $25,000 and the implication follows that size of income *is* proof of ability or obligation to pay. The amount involved is relatively small, the extra tax being only $250, but the intention of Congress in this matter is hard to understand.

It almost seems that a better course would have been to relieve corporations from all income taxes and to rely on individual incomes with an additional tax on the accumulated earnings of corporations and partnerships. There would be no inconsistency in relieving all corporations of income taxes, but an act whose stated purpose is to reduce and equalize taxes, that attains the result of raising the taxes of those least able to pay, is certainly out of line with present-day developments.

The new act was passed late in 1921 and some of the inconsistencies may possibly be traced to the haste which marked the final passage and approval. There are indications that there will be a new act or at least important changes within a comparatively short time.

The subject of tax-exempt securities has been discussed more spiritedly than ever during the past year, and the opportunity that municipal and state bonds offer for profitable investment has been advanced as a potent cause of the present depression in the industrial field. As proof of this contention, it has been pointed out that an investor, whose income is such that he must pay 50% or even less on the last increment, could more profitably purchase municipal bonds bearing 5% and hence be allowed to keep the whole 5% than to purchase industrial bonds bearing 8% and turn half over to the government. Further, that the drain of capital from industry is being accentuated by the very ease with which municipalities are enabled to raise funds, in that they are encouraged to undertake unwarranted expenditures.

Whether or not there is any weight to this objection — and there seems to be considerable — the existence of a vast amount of tax-exempt securities thwarts the purpose of our present system of federal taxation and the constitutional protection which they enjoy should be and doubtless will be removed.

University of Michigan

CERTAIN PROBLEMS OF LABOR AND THE PUBLIC ATTITUDE

HERBERT F. TAGGART

INTRODUCTION

In discussing this subject I shall first define the problems of which I shall treat. Then we shall find out who the public are and what their actual attitude is. An attempt will be made to show why their attitude is as it is, and finally, what it ought to be. In this last problem we shall consider the claims and counter-claims of those primarily interested in the problems under discussion and also what cognizance is to be taken of the principles of theoretical economics which bear on the subject.

THE PROBLEMS

The problems to be considered are principally those arising in connection with labor unions, their hopes and aims and methods of arriving at their ends. The unions themselves and their desire for higher wages, shorter hours, and better conditions present problems which must be solved, in spite of their difficulty. The insistence of the unions that their methods of action must be left open to them is disconcerting in view of the fact that these methods are usually contrary to established traditions of law and public welfare. Strikes, sabotage, curtailment of output, organization for collective bargaining, and opposition to advances in machinery and methods of production are all, according to our ordinary standards, anti-social, dangerous to the public peace and to the right of private property, and at variance with well-established economic principles.

WHO CONSTITUTE THE PUBLIC?

The question as to who constitute the public is easily enough answered by naming certain classes of people who belong to the group whose opinions we are to examine. Doctors, farmers, lawyers, preachers, small tradesmen, teachers — in a word, all ordinary, middle-class citizens who are not by birth or occupation partisans of either side.

It may be pertinent to inquire what their interest in the problems above set forth is, and why we should pay any heed to their views. Their interest arises primarily from the fact that any solution or failure to solve these problems directly affects their pocketbooks and secondarily from the fact that, being men, nothing human is alien to them, and if anything is wrong, they are anxious to have justice done. The importance of their views lies in the fact that they constitute a very large portion of society, and, in the long run, as we who are members, at least like to believe, the most influential portion. They must cast the deciding vote, if labor and capital are to be deadlocked over their differences. Hence the necessity of determining what their views are, why they are as they are, and what they ought to be.

THE VIEWS OF THE PUBLIC

An adumbration of the views of the public has already appeared in the statement of the problems with which we are to deal. The general attitude toward labor unions and their methods is distinctly unfavorable. Many of the announced aims of the unions they endorse. Justice for the workingman, democracy in industry, a living wage, and other union ideals they agree with. They are nonplussed, however, to find that apparently the methods by which the unions hope to gain these most desirable ends carry along with them other consequences which are so undesirable as largely to offset the expected gains.

Specifically, their complaints are somewhat as follows. The union is a monoply. It seeks to tax society by exercise of its powers in order that its members may profit. Its spirit is therefore contrary to the freedom of competition upon which the present order is founded. This, fully, of course, only for the more erudite of our public.

Strikes are warfare. As a method of settling industrial disputes they compare not too favorably with war as a method of settling international disputes. They involve many times violence, destruction of life and property, and suffering, not only to those directly engaged in the struggle, but to non-combatants as well. They are an industrial waste, and cannot but be reflected in higher prices for products.

Sabotage is guerrilla warfare. It, too, is an industrial waste. It is destructive to property and curtails production. It is immoral. Nothing too strong can be said in its condemnation.

Curtailment of output as a result of union rules as to hours and tasks is an especially sore point. Almost any one of us would be willing to give the laborer a fair day's pay, if only he were willing to render in return a fair day's work. And there seems something very wrong about a body of rules which undertake to tell a man how much work he may do in a day, especially when he is capable of doing a great deal more. From the point of view of society, surely, a practice whose direct effect is a reduction in the products necessary for society's well-being is hardly one to be commended.

Collective bargaining combines the opprobrium of monopoly with the injustice of the union rules, and a special grievance of the employers in that they are forced to deal with persons not in their employ or subject to their discipline in the hiring, dismissing, and making of working regulations for their men. The general public is much inclined to sympathize with the employer on this count.

WHY DO THE PUBLIC THINK THUS?

Why do the public think thus? A number of answers may be proposed.

One is tradition and training. For a number of reasons, principally historical, the capitalistic or employer's philosophy is held by the majority of the middle class. That this is the case is amply evidenced by the fact that the views just set forth are in essence the same as those of the employing class on the subjects. In one thing do the two classes differ, however, and this is fundamental. The public is inclining to agree that many classes of laborers have real grievances and that the effort to alleviate these is worthily motivated. This the employers are inclined to deny. What the public does object to is the methods used by the unions to gain their ends and right their wrongs.

Another answer is the newspaper. This is a most potent leader of public opinion. Newspaper policies are determined, immediately at least, by editors. These men belong to the group under discussion. They are representative members of the general public. As such they are nurtured in the same doctrines as the rest of their kind. Charges of capitalistic subsidy and control are made by partisans of the cause of labor, but although doubtless true in individual cases, this is hardly to be taken as a general rule. The fact is simply that the editors are, as it were, born prejudiced. They merely spread the opinions of those who do have opinions to those who do not, thus serving to unify the attitude of their fellows.

A fundamental answer is the deep-seated conviction that laborers ought to act like other civilized people and carry their grievances to the proper authorities to be redressed. Legislatures and courts supply means for private individuals to settle their disputes. Why cannot the two parties to industrial disputes depend on these same agencies? Moreover, the right of free contract alone would seem to be enough to preserve them from too great wrongs.

WHAT OUGHT THE PUBLIC ATTITUDE TO BE?

In discussing the question of the proper attitude for the public to take on these topics, it is well to realize that their present attitude is very hazy and ill-formed. The opinions just expressed are held with definiteness by only a fraction of the group. Therefore it is possible to form public opinion at the present time by an educational process; the important consideration is that it must be based on a real knowledge of the facts.

At least three distinct groups make a bid to be heard in the formation of public opinion. Laborers and employers present their sides in a most vigorous, although not unbiased, manner. The third group consists of economists whose interest in the subject is purely scientific and therefore, as nearly as may be, unbiased. The contributions of each of these groups will be taken up in order.

The unions justify their existence and their methods in a number of ways. One thing they stress is the lack of other means and safeguards. Theirs are new problems, and the old ways will not solve them. They insist that courts are too conservative, antiquated, and slow-moving, and that legislatures are not to be trusted to make good laws, even if these laws, when made, should be properly administered. They work for good legislation and make fights in the courts, but in the meantime they say that their own methods of handling the situation must be left to them. Freedom of contract they criticize as purely negative. What is needed is a positive assurance that the ordinary worker will really have a voice in making the terms under which he is to be employed. John Stuart Mill remarks that "in the case of children, freedom of contract is but another name for freedom of coercion." And this statement, the unionists claim, applies as well to the unsupported, individual workman who is confronting the modern industrial organization. Lacking waiting power, hard pressed by the competition of his fellows, unable to move to a more favorable locality, he is forced to accept whatever terms the employer offers him. Collective bargaining, then, and the union organization are necessary to insure the worker decent working conditions and fair wages. They enforce their arguments by pointing to cases, such as that of the Chicago teamsters, where the union has been the means of lifting its beneficiaries out of a veritable industrial hell.

Such instruments as strikes and sabotage are afforded a pragmatic justification in that they are the only way in which the unions are able to make their power felt. Moreover, employers are wont to do very similar things for similar reasons, with no condemnation. A few weeks ago [1] we read that certain coal mines were closing down because of slack business, thus throwing some eight thousand men out of work. The effect of such action will be, of course, a lessening in the production of coal, as well as a hardship on the miners, yet if the same diminution in production were to be brought about by a strike, it would be counted a grievous wrong to society, whereas in the other form it causes but a passing comment. Yet in striking the miners are doing no more than the employers in shutting down the mines, namely, indicating their dissatisfaction with the reward received for their efforts.

The matter of the standardization and shortening of the daily task is of sufficient importance to warrant some discussion. The unions seem to think that a consideration of the reasons for this practice may lead to a modified conclusion as to its ethical status.

This limitation of output may be due to the idea that there is just so much work to be done. And before this theory is scoffed at, it is well to note the fact that it is always a particular industry or situation at which the union rule-makers are looking. The idea is that nobody wants to leave his own trade or location in order to find the job that economists tell us is lurking just around the corner. Over the field as a whole there may be almost an infinite amount of work, but in the brick-laying trade in Ann Arbor the number of jobs is pretty definitely limited. So the daily task is shortened to bring it about that those who are at work may stay on the job all the time. The coal miners' demand for a six-hour day is a case in point. The object of that request is not so much that the men may not have to labor so long in any one day, though that would doubtless have its advantages, but that soft coal mining may cease to be a seasonal industry. And the advantage of this, not only to the miners but to society at large, is not to be gainsaid. Incidentally, this line of reasoning would go to show also that there probably

[1] This paper was read March 30, 1922.

would be no productive loss involved in this case. The smaller number of hours per day would be made up by the greater number of days per year.

In other cases limitation of hours or output is directed against the greedy employer who gauges the piece rate by the amount of work the best workmen can do. When these men begin to earn something more than a living wage he cuts down the rate for all, neglecting the fact that the majority of his men are scarcely able to maintain their standards at the old rate.

Likewise, in the case of the union opposition to the introduction of machinery and methods which displaced hand-workmen. However advantageous to society and even to laborers such procedure may be in the long run, they point out that the immediate effect is pretty hard on those most intimately concerned.

So much, in brief, for the laborers' side of the case. It will be seen that at best the arguments are only pragmatic justifications for the practices in question, and fail to reach the heart of the matter. The laborer insists that he be permitted his methods because they work and no others are open to him. The fact that these methods result in social detriment and, in the long run, to the detriment of the laboring classes themselves, is not changed. The aim of the unions is the *status quo* in industry, while social welfare demands progress.

It is to be remembered also that complete domination of industry by the laborers would be fully as unfair to the employers as is the present domination by the employers unfair to the workers. Some curious results might be expected to follow such domination. In the printing trades, where the unions are now very powerful, we have the spectacle of hand-workers duplicating the same tasks in the same shops as are done by the electrotyping process. Such waste and hindrance of progress are characteristic of present-day union policies.

For these reasons we cannot accept in their entirety the union arguments. A union autocracy seems, in the long run, to have few advantages to offer over an employers' autocracy, even to the laborers themselves. A serious practical difficulty which would confront us in adopting the union position is the fact that the union philosophy is essentially a group philosophy, and not a social one. It does not even include all laborers, and in addition we have the spectacle of internecine warfare in the ranks of the unions themselves. It is not necessarily to the advantage of any one union group that any other should be benefited. So that we should by no means be assured of peace even if we yielded to the unions completely.

The specific arguments of the employers against the union program have already been given under the head of the present attitude of the public. It is an unfortunate but not unaccountable fact that the man of the street, for the most part, knows only one side of the question. Employers have always taken it for granted that their interests were society's interests, and most of us have taken them at their word without further inquiry.

Yet nothing is more certain than that the interests of employers and society clash at certain vital points. Few valid arguments can be made against the proposition that it is to society's interest to protect women and children from work that stunts them mentally, morally, and physically. Yet it is to the interest of the employer who can make use of such labor to do so, because he can get it cheaply and thereby outstrip his competitors. At first glance it is hard to see how it is just that employers should be interfered with in the conduct of their affairs, but when one considers such conditions as are found in sweat shops, doubts begin to arise.

Another example of the conflict between the interests of society and the employer, though not in the field of labor, will serve to bring out the cause of nearly all such discrepancies. In the telephone industry there have been many inventions which, if applied, would revolutionize the service, making it far more efficient and satisfactory. But many of these discoveries are not made use of, simply because the companies cannot afford to do it. In order to do it they would have to scrap enormous amounts of their present equipment, and in any reasonably brief period of time they could never be repaid for this cost, nor could they, even if they were not regulated as to their charges by public authority. Here, then, we see society cheated out of a great blessing by the mortality of its members. Hence we arrive at the conclusion that laborers are not the only ones who block progress.

We conclude that an unbridled employing class would be as disadvantageous as the other extreme. Things that have happened with no regulation, either by the public or by a powerful labor group, such as the evils already mentioned, show us that here, at least, is a point at which self-interest cannot be entirely trusted. Not being able to settle the question in favor of either the employers or the employees, we turn to the economist to see what he has to offer.

The economist, we find, lives in an ideal world, composed of perfect economic men, perfectly motivated by self-interest, who offer their wares in perfect markets where perfect competition prevails. Evidently, then, to apply his laws to society as it actually exists, we must take into consideration the many imperfections which are seen to exist in these various particulars. We have to distinguish carefully between the theoretical present order and the actual present order.

The economist tells us, for instance, that the supply of jobs is not absolutely limited. Yet we find, at any given moment, in normal times, a million men capable of work, but out of employment. We learn that competition of employers will keep wages from falling to the level of the meanest-spirited workman. Again we are confronted with the idle million. It is a question, perhaps, as to whether this fact is sufficient seriously to affect the validity of the principles: there are numerous reasons why men become out of work, but at least we see that things do not function as smoothly as they might.

The economist looks at things from a social, which is necessarily a long-time, standpoint. The individual laborer and employer must regard his problems from an individual, or at best a group, standpoint, and this is necessarily short-time. The economist tells us that low wages will automatically correct themselves. But this process takes such a time as the individual is unable to wait. We are assured that capital does not compete with labor; that it is, in fact, an adjunct to labor, and the more there is of it the better off is everybody, including laborers. But the skilled hand-worker who is displaced by a machine knows better. The pitiful struggles of hand-weavers and spinners to compete with power machinery in the days of the Industrial Revolution sufficiently illustrate the point.

Thus we have a number of disturbing factors which force us to qualify our application of economic principles to the problems at hand. Among these are the immobility of labor, industrial friction, the working of motives other than that of pure self-interest, and, very important for our purposes, the contrast between the short- and long-time viewpoints.

Of what use, then, are the principles to us in our search for a proper attitude for the public? J. S. Mill gives us a clue. "Many persons," he says, "mistake permanent laws for temporary accidents arising from the existing constitution of society, which those who would frame a new system of social arrangements are at liberty to disregard." The economist, then, can tell us how far we can go in certain directions in alleviating social wrongs, and what the consequences of our acts will be. A fine example of this is Professor Taussig's warning in the case of certain contemplated minimum wage requirements that a falling off in the demand for that type of labor would be the inevitable result. Conversely, Mill says, "temporary accidents" are often confounded with "permanent laws." The economist can save us from that. No principle is more important in its practical consequences than the one concerning the harmonious coöperation of capital and labor. Yet were we to accept the labor union viewpoint we should consider the case to be precisely otherwise, and, acting on that assumption, should do society a great injustice, for it is an accepted rule that in general a maximum of production is desirable.

This discussion could be expanded indefinitely, but enough has been adduced to indicate that, though we must qualify our application of economic principles, we must nevertheless be guided by them.

Now let us set down the conclusions which the ordinary

citizen might arrive at when taking all these claims and counterclaims into consideration.

He should conclude that the laborer has some valid grievances. Some of these have been mentioned. Very possibly any worker who is not receiving enough pay to sustain him and his family at a reasonably high standard has a grievance. But the methods taken by the laborers to remedy their difficulties are not good. They are not those which will work, in the long run, nor are they, certainly, such as will bring about a harmony of interest among all classes of society. But the workers are not utterly to be condemned for resorting to these methods. They are driven to them, in general, because of the lack of others which would be as effective. Political and judicial instrumentalities are too slow and unsympathetic, and the working out of economic laws takes too long a time.

Nor does the employer have all the wrong of the matter. He, too, is a victim of circumstance, not so much, to be sure, as he would like us to believe, but still largely unable to help himself. He has rights in the premises which must be respected.

The economist may safely be trusted to give us long-run, guiding principles, but must not be taken too hastily or literally.

A middle ground, to some extent a compromise, is strongly indicated. Its general principles have been given. Justice is its ideal. The practical applications would have to be worked out by trial and error. There is confessedly no royal road which we may travel. Just what a given laborer ought to get must be determined by taking many things into consideration. His needs, as a human being, ought probably to be given more direct weight in the final result than has often been the case. It seems to be a possibility that we may get some goods too cheaply, on account of the human cost involved in their production. The consensus of opinion has endorsed this view as to white phosphorus matches and the products of child labor. It may well be true in many other cases.

Some faint glimmerings of the dawn of a new industrial day can be discerned by the optimistic observer. Attempts at the eradication of fluctuations in employment and occupational accidents and diseases are encouraging signs. Here and there we find situations in which employer and employee are honestly trying to coöperate to the end that both shall receive what they deserve. In this the public scarcely stands to suffer. And many employers, paying far greater wages than ever before, and taking an interest in their employees as human beings, and not as mere sources of labor power, have found to their astonishment that they are making it pay, and pay handsomely.

University of Kansas
Lawrence, Kansas

LIPSANOCYSTIS TRAVERSENSIS, A NEW CYSTID FROM THE DEVONIAN OF MICHIGAN

G. M. EHLERS AND J. B. LEIGHLY

In the Museum of Geology of the University of Michigan there was recently found a cystid with a label, in Dr. Carl Rominger's handwriting, bearing the name *Lepadocrinus hamiltonensis* Rominger, and indicating that the specimen was collected at Partridge Point, Thunder Bay, Michigan. This cystid is probably the one that Rominger had in mind,[1] when he recorded the occurrence of the genus *Lepadocrinus* along with several genera and species of corals, echinoderms, bryozoa, brachiopods, and trilobites in the rocks of Partridge Point, which he regarded as belonging to the Hamilton group. No description, however, of this cystid, or of the species *Lepadocrinus hamiltonensis* Rominger, occurs in the literature. It seems very probable that Rominger, after recording this occurrence of *Lepadocrinus*, thought the cystid to be a new species, labeling it *Lepadocrinus hamiltonensis*, but never publishing a description of it.

Upon examination, it became evident that this cystid did not belong to the genus *Lepadocrinus* (= *Lepocrinites*), but to an undescribed genus, for which the name Lipsanocystis is proposed.

Class CYSTOIDEA
Order DICHOPORITA
Family CALLOCYSTIDAE
Subfamily Apiocystinae

Lipsanocystis, n. gen.

Definition. — Apiocystinae with 20 plates, arranged as follows:

Plates, 4, 1, 2, 3, in basal row;

Plates 5, 6, 7, 8, 9, in second row;

[1] Rominger, Dr. Carl, *Geology of the Lower Peninsula, Geological Survey of Michigan*, 3: Part I. 41. 1876.

Plates 10, 11, 12, 13, 14, (15), in third row;
Plates 16, 17, 18, (13), 19, 15, in fourth row;
Deltoid 23 double, in fifth row.

Deltoid 23 is double, and contains both the madreporite and the hydropore. The former is divided, and appears as two elevations, each with an orifice in its apex. The hydropore appears as a small circular opening below and between the two parts of the divided madreporite.

Anal opening moderately large, entirely enclosed by plate 13, except for about 2 mm. of its periphery, which is formed by the upper edge of plate 8. One basal and 2 upper pectinirhombs, with numerous dichopores, situated respectively on plates 1 and 5, 14 and 15, and 12 and 18. The halves of the pectinirhombs on plates 5, 14, and 12 surrounded by conspicuous raised margins, those on 1, 15, and 18 without marginal walls.

Ambulacra four, simple, extending to the basal row of plates. These are RI, RII, RIV, and RV. Brachioles relatively few in number and rather widely separated.

Column unknown.

Genotype, *L. traversensis*. No other species known.

Lipsanocystis shows in its structure relationship with other Apiocystinae, especially *Apiocystites* and *Tetracystis*. Dr. Charles Schuchert,[2] who has examined the specimen, pronounces it "a changed Apiocystites, having become more depressed and therefore has changed somewhat the arrangement of the plates." It has pectinirhombs resembling somewhat those of *Tetracystis chrysalis* Schuchert, but differs from all other genera of the family in having the anal opening almost entirely enclosed by plate 13, only a small part of the border being formed by plate 8. (See Text Fig. 5.) Since this genus is apparently a holdover of a Silurian type, the name *Lipsanocystis* is given to it.

[2] The writers wish to express their appreciation of the kindness of Dr. Schuchert in examining the fossil and offering valuable suggestions as to its structure and relationships.

Lipsanocystis traversensis n. sp.

(Plate X, Figs. 1–5, Text Figs. 4, 5, 6)

Greatest length of theca of holotype 22 mm., greatest width 19 mm., thickness 15 mm. Form and appearance of entire

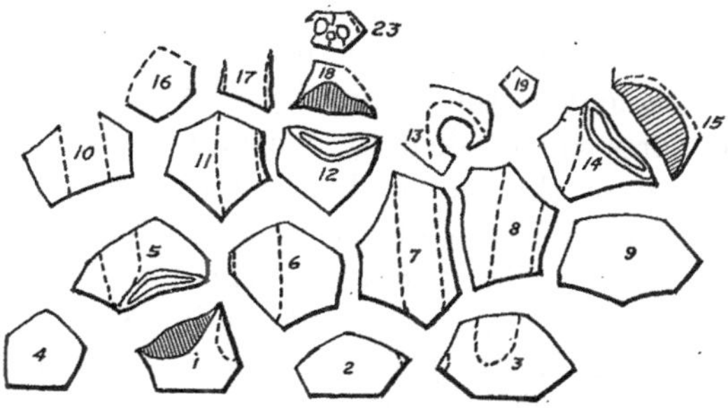

Fig. 4. Diagram of thecal plates.

theca shown in Plate X, shape of individual plates in diagram, Fig. 4.

Ambulacra prominent, occupying extremely shallow grooves

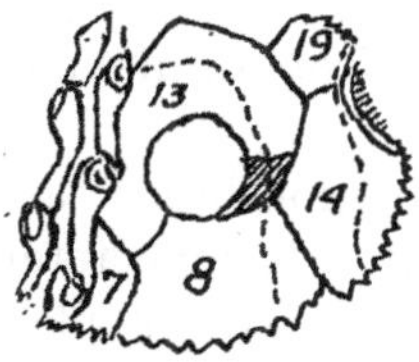

Fig. 5. Sketch of anal side, showing relationship of anal opening to adjacent plates. ×2.

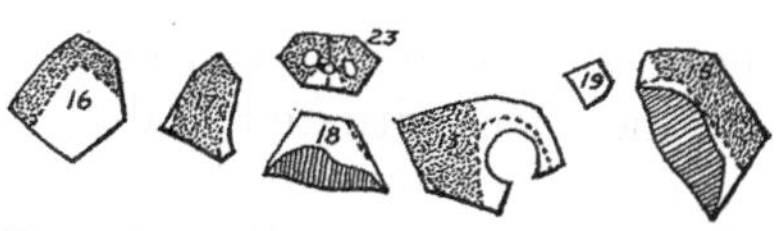

Fig. 6. Diagram showing probable outlines of plates composing rows 4 and 5. Parts covered by ambulacral plates stippled.

in the thecal plates. Brachioles 14 to each ambulacrum, or 56 to the entire theca. Length unknown.

Pectinirhomb on plates 1 and 5 has about 45 dichopores, that on plates 12 and 18 about 42, that on plates 14 and 15 about 50.

Anal pyramid unknown.

Column unknown.

The single specimen, No. 5414, Museum of Geology, University of Michigan, upon which this description is based, is from the upper part of the Thunder Bay division of the Traverse formation, at Partridge Point, Alpena County, Michigan. Since the name Traverse is now applied to the formation from which this fossil was collected, the specific name *traversensis* is used instead of Rominger's *hamiltonensis*.

The description above is based upon a single specimen in the Museum of Geology of the University of Michigan. The details of the oral parts are masked by the ambulacra. A conjecture as to the form of the plates thus partly covered is shown in Fig. 6, in which the parts of the plates covered by the ambulacrals are stippled. The two upper circles on deltoid 23 in this figure are interpreted as a divided madreporite, with the hydropore appearing faintly just below and between the parts of the madreporite. These latter parts are elevated, and in appearance closely resemble the adjacent bases of brachioles. They are connected by an elevation above the surface of the plate, and are interpreted as the madreporite on the basis of their apparent similarity to the structure described by Schuchert[3] in *Callocystites jewetti:* "Madreporite ∞-shaped and situated on the two parts of plate 23. Hydropore immediately beneath the madreporite, very small."

There is some doubt concerning the relationship of a small part of the theca situated orally from plate 15, and anteriorly from plate 19. This may be interpreted as a part of plate 19, a part of plate 15, or perhaps as deltoid 20.

Another doubtful detail is the form of plates 10, 13, 15, 16, 17, and 23. The hypothetical completion of the plates in Fig. 6 is based upon the apparent position of the angles of the plates as indicated by the intersection of the visible parts of their sides produced. The doubt concerning these structural details can only be resolved by the examination of additional material.

University of Michigan

[3] Schuchert, Charles. *On Siluric and Devonic Cystidea and Camarocrinus, Smithsonian Miscellaneous Collections,* 47: Part II. 244. 1904.

EXPLANATION OF PLATE X

Lipsanocystis traversensis, n. sp. (Natural size)

FIG. 1. View of anal side of theca. Upper part of ambulacrum RV missing.

FIG. 2. View of antanal side.

FIG. 3. Oral view of theca. Deltoid 23 with divided madreporite appears immediately at left of mouth opening.

FIG. 4. View showing the upper left-hand pectinirhomb and the configuration of the ambulacra. Anal side to the right.

FIG. 5. Opposite side to that shown in Fig. 4, showing upper right-hand pectinirhomb and fragment of column adhering to basal plates.

PLATE X

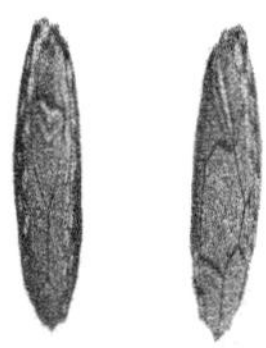

FIG. 1

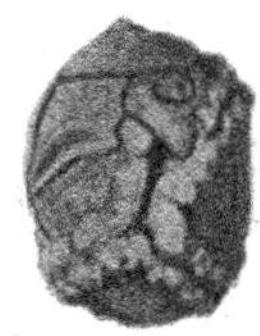

FIG. 2

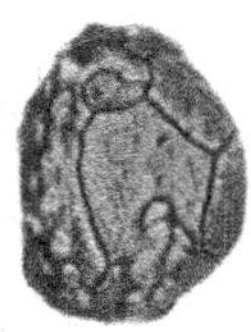

FIG. 3

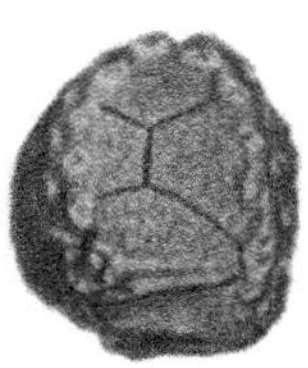

FIG. 4

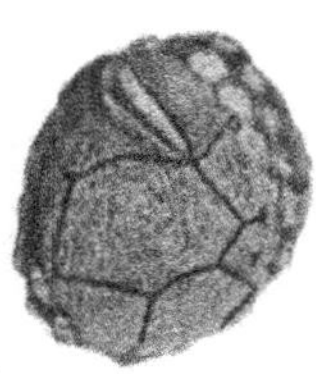

FIG. 5

Lipsanocystis traversensis, n. sp.

METEOROLOGICAL SUMMARY, DOUGLAS LAKE, MICHIGAN

FRANK C. GATES

In the table following is recorded a summary of the meteorological data obtained at the University of Michigan Biological Station on Douglas Lake, Cheboygan County, Michigan, for the ten-year period, 1912–1921.

The table gives the data for periods of approximately one fifth of a month during the two months in which the Biological Station is in session. The exponential figures indicate the number of days in each period which are missing from the record. As the Station has closed about August 24, the records for the last week of August are very scanty. Temperature is expressed in degrees Fahrenheit and precipitation in inches for the periods indicated. In each case the averages are obtained from the grand totals and expressed to the nearest tenth of a degree or one-hundredth of an inch.

METEOROLOGICAL AVERAGES, 1912–1921

		Average Maximum	Average Minimum	Mean	Precipitation
July	1–6	80.3^{3}	55.7^{5}	68.0	$.40^{3}$
	7–12	79.4	55.7	67.6	.34
	13–18	79.5	57.6	68.6	.22
	19–24	81.7	57.9	69.8	.31
	25–31	81.9	59.7	70.8	.51
Average		80.6	57.4	69.0	1.79
August	1–6	77.1	55.4	66.3	.44
	7–12	76.9	57.2	67.1	1.27
	13–18	76.8	56.1	66.5	.61
	19–24	76.6^{12}	56.1^{10}	66.4	$.49^{10}$
	25–31	73.4^{60}	55.3^{56}	64.4	$.16+^{56}$
Average		76.7	56.1	66.4	2.90+

KANSAS STATE AGRICULTURAL COLLEGE
MANHATTAN, KANSAS

METEOROLOGICAL DATA, DOUGLAS LAKE, MICHIGAN

FRANK C. GATES

In the notes and table following is recorded a summary of the meteorological data obtained at the University of Michigan Biological Station on Douglas Lake, Cheboygan County, Michigan, during the summers of 1920 and 1921.

July, 1920. — Following a rather warm June, the month of July was the coldest July in the history of the Station and lacked only 1.5° of being colder than any August as well. It was particularly cold during the first three weeks, although during that time no new minimum temperatures were established. Precipitation during the month was considerably above normal, although July, 1915, had two more days of rain and July, 1913, had .3 of an inch more rain.

August, 1920. — This month was characterized by slight departures from the normal; the temperature was a little below average and the rainfall a little above. A new absolute minimum temperature, namely 40°, was, however, established on the 23rd.

June, 1921. — After a sharp, killing frost early in June, the month was very warm and dry. During this month vegetation proceeded very much farther than usual, particularly in the case of the aquatic plants, many of which were four to five weeks earlier than usual in coming into flower.

July, 1921. — July established itself as the hottest July of the Station record, although the absolute maximum temperature of 104° was not reached. The rainfall was somewhat above average.

August, 1921. — The season changed abruptly in August, the low temperatures of which resulted in its being the coldest August with the exception of that of 1912. It was likewise the rainiest August with the same exception. Before the onset of the rainy weather, the lake had receded to the lowest level in the history of the Station. According to old settlers in the vicinity, it was the lowest it had ever been in over thirty-five years. The water also reached the highest temperature that has been recorded at the Station, 98° on July 8. It passed over 75° on June 23 and did not fall below 77° until July 28, during part of which time, June 29 to July 18, it was continuously above 80°. The maximum water temperature previously noted had been 86°, but water temperatures above 80° had been noted only four or five times in the history of the Station. (Water temperatures during 1921 were taken twice daily by Mr. C. L. Hubbs.)

The table following shows the summaries for the seasons of 1920 and 1921; temperatures are expressed in degrees Fahrenheit and precipitation in inches.

METEOROLOGICAL SUMMARY FOR SUMMER OF 1920

Month Year Days of record.	June 1920 22	July 1920 31	Departure	August 1920 25	Departure
Absolute maximum	96	87	–17	88	–11
Average maximum	78.3	73.1	– 7.5	76.4	– 0.3
Absolute minimum	43	39.5	+ 1.	40	0
Average minimum	56.2	52.4	– 5.0	53.0	– 3.1
Mean temperature	67.3	62.7	– 6.3	64.7	– 1.7
Precipitation	2.34	2.77	+ 0.98	3.54	+ 0.64
Days of precipitation	10	11	+ 2	6	– 2

METEOROLOGICAL SUMMARY FOR SUMMER OF 1921

Month Year Days of record	June 1921 24	July 1921 31	Departure	August 1921 28.5	Departure
Absolute maximum	91	100	– 4	87	–12
Average maximum	82.6	87.6	+ 7.0	72.4	– 4.3
Absolute minimum	28	52	+13.5	43	+ 3
Average minimum	54.6	62.8	+ 5.4	54.2	– 1.9
Mean temperature	68.6	75.2	+ 6.2	63.3	– 3.1
Precipitation	1.11	2.41	+ 0.62	4.13	+ 1.23
Days of precipitation	5	9	0	11	+ 3

KANSAS STATE AGRICULTURAL COLLEGE
MANHATTAN, KANSAS

ILLUSIONS OF ORIENTATION

NATHAN A. HARVEY

It is a very uncomfortable experience when a man's consciousness and the compass do not agree; when, for instance, he has a feeling that a certain direction is east and the compass says that it is north. It is an experience that is very common, although some persons never have it. An attempt to discover the number who do leads to the conclusion that from one third to one half of all persons are susceptible to it. There are persons, however, who seem to be entirely free from confusion about directions.

Although the experiences of being turned around are very common, but little attention has been given to their description or explanation. The standard or classical paper upon the subject is Binet's *Reverse Illusions of Orientation*, translated and published in the *Psychological Review*, Volume I. In this paper, Binet describes nine cases, and discusses the phenomena, although not in a very exhaustive way. Since Binet's article was published, there has been comparatively little investigation of the subject.

The present paper is the result of the collection and study of quite a large number of cases, which present diverse variations in the phenomenon, as well as the result of various experiments which I have conducted upon myself. It will be well for us to examine some typical cases.

Number 1 is the case of Miss Grace E. Miss E. reports that she is always turned around in Battle Creek and in Kalamazoo. At her own home, and in Ypsilanti, north seems north and east is east, but in both Battle Creek and in Kalamazoo, east seems to be north, and the sun seems to her to rise in the north every morning.

Here we have a typical example. Although Miss E. knows,

as an intellectual fact, which way is north, and can indicate the direction the compass needle will point, she has a deeply seated feeling that both the sun and the compass are wrong in their indications. It is this feeling, of which she is intensely conscious, that constitutes the illusion.

It will be noticed also, that the amount of turning is 90 degrees. In nearly every case, the amount of turning is exactly 90, or 180 degrees.

A third noticeable characteristic is that the error is the same in direction in both places. We shall find by a study of many cases that the direction and the amount are the same in any one person for much the larger number of experiences. In my own experience, whenever I am turned around, in about four cases out of five the turning is of such a nature that east is experienced as north, and the sun seems to rise in that point.

A fourth characteristic which is represented by the case of Miss E. is that, in the same locality, the error in direction is always the same. There is no variation in the false direction in a particular locality.

But a collection and examination of many cases will show that no single one of these characteristics of the phenomenon is invariable, nor the same for all persons. As an example of a variable illusion, we may study the case of Miss Alice F. Miss F. lives several miles from the nearest town, and has frequent occasion to visit that place. She always goes over the same road, but sometimes it appears to her that the road is leading west, and again that it is going south. Sometimes it appears to her that she is going neither south nor west, but southwest, in a direction between the two. So in this case, we have an example of a deviation from the ordinary in two characteristics; the direction is variable and the amount of the shifting is either 90 or 135 degrees.

After a false direction has become fixed in a person's consciousness, there is little probabil'ty that it can ever be subsequently corrected. This fact gives rise to some very curious and interesting experiences. The change from a false system of directions to a true one, or from a true system to a previously experienced false system, occurs instantaneously, or very nearly so. An experiment by the writer will illustrate this point.

Champaign, Illinois, is one place, in which, according to my experience, the sun always rises in the north. In July, 1920, having occasion to travel through Champaign on the train over the line of the Illinois Central Railroad, whose general direction there is south, I was interested to see if the directions would be the same as I had experienced them to be some fifteen or eighteen years before. The train was headed south until we were very close to the station. I was seated with watch and notebook in hand ready to record the experiences that might occur. Just as we approached the station, the thing seemed to hang in the balance for several seconds, and then it swung around, and we were going east, instead of south, as we had been five seconds before. Then I waited to see if the original direction would be resumed after we had passed the town. When we about two miles out of town, the universe, without any hesitating delay this time, instantaneously seemed to swing back, and we were again headed south, as we had been before passing the station.

The influence of the method of learning directions has not been sufficiently considered. Many children get their first knowledge of directions from the study of a map at school.

The experience of Miss Geneva S. is very instructive upon this point. Miss S. describes her sensation by saying that she is permanently turned around. She is never in a place in which the directions seem to her to be what other persons say they are. To her, north is always south, and east is always west, and she cannot make them seem to be any other way. She attributes this feeling to the fact that she learned her directions at school from a map that was hung up on the south side of the room. She expresses her opinion that a teacher who teaches directions from a map hung on the south side of the room ought to be hung on the north side herself.

Very similar to this case is that of Miss Mae J. which was reported to me only last week.[1] Miss J. is not turned around

[1] This paper was read March 30, 1922.

in her own town, but in other places she is always turned around. That is, in her home town, or in Ypsilanti, or in any other place that she visits in person, the directions are in accord with the compass. But whenever she has to think of some place which she has learned from a map, she is turned around. Whenever she is asked to point toward California, she feels that she must point in the direction that they tell her is north. In the same way, Canada is east, Boston is south, Florida is west. She believes that this very uncomfortable experience arises from the fact that she learned the directions from a map that was hung up on the east side of the room.

But there are many persons who are not troubled at all by directions. They never have any consciousness of direction, and are not in the least troubled by the lack. This feeling is well expressed by the account of Miss Mabel M. S. Her account is given in her own words: "I get turned around and have mighty little sense of compass directions. In fact, I could not begin to tell the directions in a different city, and I don't care to be bothered with them, for they only confuse me. At home, south is always north, and vice versa. East is always away from the sun. That is all I know about directions, and I never use them. I only get confused when I try. If any one tells me, in response to my inquiry, to go so many blocks north and then east, I am simply at sea and don't know where to go. If, however, a person should point in the direction that I should go, and then designates a turn to the right or to the left, and indicates certain landmarks, such as a building or a bridge, I am all right. The best set of directions that I have ever had was given to me in Philadelphia last summer, when I was directed to go 'kitty corner two blocks down and three blocks up.'"

The completeness of the illusion is represented by the example of Mrs. Ruth B., who reports the experience of her husband's mother. It is a case of apparently pathological orientation, although no definite pathological symptoms are indicated in the account. The experiences have occurred as frequently as every three or four months for the past fifteen years. She describes them in about the following manner: When she gets up in the morning the house appears to be on the other side of the street from what it was the day before. The other directions are all wrong. When she decides to go to the kitchen, the directions are so much out of the way that she has difficulty in reaching the kitchen at all. She is lost in her own house, and is unable readily to find her way around it. She feels as if she is in a strange place, and gets into the wrong room as frequently as she does the one into which she wishes to go.

She is unable properly to locate the neighbors' houses and cannot find her way around the neighborhood without a great deal of trouble. She knows exactly what the difficulty is, but it is impossible for her to make things seem straight. The experience will last for three or four days, when everything turns back to its former state, and seems to be familiar, with the same system of orientation as it was before.

It is probable that this account ought to be referred to the category of double, or alternating, personality, rather than to that of illusions of orientation. The illusion of orientation seems rather incidental to the deeper and more complex series of phenomena known as double personality.

I myself am exceedingly susceptible to directions. I always have a notion, accurate or not, that a particular direction, in any locality in which I may be placed, is north, and the other cardinal points are correspondingly located. This disposition is so pronounced that in every dream which I have, the directions are as evident as they are in my waking experiences. I am unable to remember any dream involving spatial elements in which the cardinal directions were not evident. I believe this is a feature of orientation that has not previously been noticed.

Many attempts to relieve myself of the uncomfortable feeling associated with being turned around have been made, but always without success until about two years ago. As these experiments are indicative of a satisfactory theory of orientation, it is well to detail them at considerable length.

About two years ago, I was going by trolley from Grand Rapids to South Haven, when I discovered, by observing the sun, that I was turned around, and was really going west when

I felt myself to be going south. Then as an experiment, I tried to think of a place along the trolley line between Wayne and Ypsilanti, where the general direction of the railroad tracks is east and west, and found that I was able by doing so to make it appear to myself that I was going west.

With this experiment in mind, I tried another. On the second day of last June, I got off the train at Alpena where the railroad track runs north and south, and the directions all appeared to me to be straight. Having some time to wait, I inspected the station building from all sides, particularly from the east. I came back to Alpena very late that night, stopping at a hotel. In the morning I was directed to go down the street a few blocks and turn toward the left. I started north, as it seemed to me, then turned west, toward my left, expecting to come to the station building from the east. But at one street-crossing I beheld the station to the north of me, when I thought it ought to be toward the west. Quickly I realized that I was turned around, and decided to try an experiment. I thought of the station as I had seen it the day before from the east, and immediately the directions seemed straight. Then I took out my watch, and for two minutes I tried to see how many times I could make the directions shift through an angle of ninety degrees. I shifted them twelve times in two minutes, and it appeared to me that I might have kept on indefinitely, although I thought I began to notice that the more recent direction, the north, was becoming less vivid.

A very similar experiment was rendered possible last August, when on leaving Pittsburgh over the Pennsylvania Railroad for Cleveland, I found myself sitting on the left side of the train, running alongside the Ohio river, which was flowing toward the east, in the same direction that the train was going. Then the sun shone somewhat through the clouds, and it became apparent that I was turned around 180 degrees. I immediately thought of a place somewhat similar in character, between Michigan City and Hammond, on the road to Chicago, and the directions became straight. Then I thought of the eastward direction as it had seemed and the thing shifted back as it had been at first. I repeated the shifting as much as a dozen times, and could make it appear the one direction or the other.

It appears, then, from these carefully conducted experiments, that the illusion of being turned around is of exactly the same class as the illusion of the duck and the rabbit, the stairway illusion, or that of the wife and mother-in-law. The best explanation that any one can give of such illusions is one in which it is assumed that when anything is perceived, a nervous impulse goes through some combination of cells in the brain. Whenever the same combination of cells is traversed, the same thing is seen. We may by a process of attention shift the impulse into a slightly different combination, which would correspond to the perception of a slightly different thing.

By some supposition such as this, we may believe that whenever we are conscious of one system of directions, a nervous impulse is traversing some combination of cells that would correspond to the perception of this orientation. When we are turned around, the impulse is traversing a somewhat different combination. We may shift the impulse from one combination of cells to another by a process of attention, and that is about as far as our explanation will permit us to go.

Summing up, then, we may notice the following points:

1. From one half to two thirds of all persons experience occasionally the phenomena of being turned around.

2. About one half to one third of all persons do not have involved in their perceptions the element of direction.

3. In some persons the element of direction is universal, appearing even in dreams.

4. Persons may be permanently turned around, if the directions have been learned from a map inappropriately hung.

5. Being turned around may appear as a feature in the phenomena of double personality.

6. Illusions of orientation belong to the same class of illusions as the stairway illusion, the duck and the rabbit, and the wife and the mother-in-law illusions.

7. The easiest and the clearest explanation of the illusion is one that considers the shifting of the nervous current from

one combination of cells to a slightly different combination, the concomitant of the transition from one system of directions to another.

8. It is possible by a process of attention so to shift the nervous impulse that the system of directions may be rapidly changed.

State Normal College
Ypsilanti, Michigan

SOME ASPECTS OF THE CHLORINATION TREATMENT AT THE FLINT FILTRATION PLANT

R. S. BUZZELL

It is not the purpose of this paper to present anything new, but rather to discuss the chlorination problems and natural tastes in connection with the water-purification treatment at Flint.

To produce a pure wholesome water from a questionable surface water and to obtain this result without objectionable tastes, is the main problem of the plant operator. The writer has a lot of sympathy with the consumer who objects to a drinking water that has so rank a taste that he cannot use it, even if he is assured that it is free from pathogenic bacteria.

It has been stated that the wide use of chlorine in water purification has become a detrimental factor in filtration plant practice, because by its use the other steps in purification, namely, proper application of chemicals, time of mixing and sedimentation, and the care of filters, can be hurried to a point of inefficiency, and yet turn over to the consumer an acceptable water. We believe chlorine should not be used, except in an emergency, to permit any slackness in the other chemical and mechanical steps in water purification. The aim of the operator should be to keep every unit in the process working to its highest efficiency.

Apparatus for the application of chlorine has kept pace with modern progress in purification work. Hypo-chlorite was used at Flint for several years, and although effective, it was more difficult to keep away from the taste produced by slight overdoses than it is today with our present plant construction and dual installation of chlorine gas machines.

Originally, the hypo solution was kept agitated by motor-driven apparatus and was applied by means of float box and set orifice. The hypo solution applied to the filtered water was not always clear, but often carried small particles of undissolved hypo which seemed to favor an overdose unless the greatest care was exercised and controlled. The solution was applied to the suction of the high-service pumps. On account of the short time interval from the point of application to the point of taking samples at the laboratory tap, it was found necessary to make tests at the first city tap, a mile or more distant. It is obvious that when too much chlorine was found at that point, it was too late to avoid taste. Although the chlorine was applied by what is known as an invariable dose, we feel that we were fortunate in that the chlorine taste did not develop more than it did. Over a period of three years, chlorine entered this range but three times.

About the time of the installation of the first chlorine gas apparatus, in 1917, a gradual increase in mineral and organic matter in the river water began to manifest itself. This change has been particularly noticeable since 1919. The color has increased from an average of 20 parts per million to an average of 50 parts per million; alkalinity has an average of 215 at present compared with 275 formerly. The hydrogen-ion value at present averages 7.2. The increase in color, which we are most interested in as pertaining to the chlorine absorption value, has given no indication of remaining at this higher constant average.

As a result of investigation for the reason of this change in the character of the river water at Flint, we have found that an increase in farm activities, improved roads, and some construction of new county drains on the watershed may be important factors in the increase of colloidal color and turbidity.

The change in color has necessitated an increased dosage of chlorine. To add chlorine to the absorption value of the filtered water, which has been an average of .46 parts per million, then to increase the amount from .12 to .15 parts per million excess, reduces gas formers to the point where the water more than meets the requirements of the Treasury standards.

In swinging into the method of applying chlorine by the

absorption method which indicated a variable dose, we were reluctant at first to rely alone on the O. Tolidin test for excess chlorine. Many operators are still clinging to the invariable dose, from fear of the criticism that would result if they overdosed the chlorine into the taste range.

The methods used at our laboratory worked out nicely. We adopted the following technic in using O. Tolidin, and worked out the starch iodide method to a point delicate enough to check and recheck the O. Tolidin test for excess chlorine. The results gave us absolute confidence in the amount of excess chlorine as we varied the dosage.

In making up the O. Tolidin reagent we find it very important to allow the solution to stand at least a week or ten days before any attempt is made to use it as an indicator. In using the O. Tolidin, one c.c. of the reagent is added to one hundred c.c. of the chlorine-treated water, and after mixing and standing seven minutes for water 50° F. or less for the development of the maximum color, twenty-five c.c are diluted to one hundred c.c. and compared to the standard colors.

The best point of chlorine application would be to the filtered water before it enters the clear well, but we obtain satisfactory results by applying it to the settled water flume leading to the sand filters. Very constant results are obtained by using one chlorine machine to the amount of the absorption value with a slight excess. This has a sterilizing action on the sand filters and clear well. Chlorine machine No. 2 can be used for application to the suction of the high-service pumps. Only a small chlorine treatment is necessary at this point and then only when the excess over the absorption value falls less than one-tenth part per million.

An obnoxious phenol taste has developed in the filtered water but once since 1917. On the afternoon of October 8, 1920, the new eight-million gallon clear well was to be connected to the old two-million gallon clear well then in use. Previous to cutting through the concrete wall for the installation of the 60″ gate-valve to connect the two reservoirs, a bulkhead had to be placed. The bulkhead, before being lowered, had been heavily coated with a tar mixture and it was lowered into position by the construction foreman who afterwards informed us what had been done. The bulkhead was ordered removed at once, but before it could be accomplished a strong phenol taste similar to that caused by an overdose of chlorine was present in the water from the laboratory tap. The only effective remedy was to overflow the clear well and wash out the floating tar oil. The taste was eliminated in the clear well in about five hours.

Complaints were numerous that evening, but a frank explanation of the incident with the assurance that the taste was harmless and would last only a few hours seemed to satisfy all inquirers. The occurrence was caused by a misunderstanding between the construction superintendent of the work and his foreman.

For several years no taste has developed from the use of chlorine in our water-supply. But during the spring flood period, from six to ten days, a grassy or, more particularly, a woody taste is prevalent, and some complaints are heard at this time. The only treatment that we have found of value in minimizing this taste is a laboratory one. Violent aëration is necessary together with a water treated with lime to a neutral point. In feeding lime in many different amounts with the plant process with waters of high turbidity and low alkalinity, if there is any reduction in the grassy taste, it is not noticeable.

The water at Flint is free from trade waste and, excepting the short spring flood period, it has naturally an agreeable and acceptable taste throughout the year.

The control of the chlorine treatment at Flint is most important, if the treated water is to be kept free from objectionable tastes. For treatment with the present color and organic content of the water, applying chlorine to an excess over the absorption value varying, according to the temperature of the water, from .05 to .22 parts per million attains the results aimed for, and only by so doing can the water be accepted by the community with the most favorable attitude.

Water Department
Flint, Michigan

A SIMPLE QUANTITATIVE PRECIPITATION REACTION FOR SYPHILIS

R. L. KAHN

This precipitation reaction is based on the employment of syphilitic serum and alcoholic extracts of heart muscle. Originally a cholesterinized antigen only was employed and the test was completed after four hours' incubation. Later in the experiments a non-cholesterinized antigen was employed as a check on the sensitive cholesterinized one, and the incubation period was extended to about sixteen hours. The antigens are prepared so as to contain a concentrated amount of antigenic substance and are diluted for the tests with approximately the smallest amount of saline which will hold them in solution. This gives the antigens marked precipitating power when in contact with positive serums.

This reaction possesses the following salient features:

1. The diluted antigens possess considerable keeping qualities, eliminating the necessity of diluting the antigens before each series of tests.

2. The strongly positive serums react, in most cases, spontaneously with the cholesterinized antigen.

3. Small quantities of serum and antigen may be employed when it is difficult or impossible to obtain larger amounts of serum.

4. The precipitates are recognized with comparative ease.

The preparation of antigen. — Fresh pig or beef heart is freed from fat, fiber and blood-vessels and passed several times through a meat chopper and dried. The dried material is broken up into small particles and passed three times through a coffee grinder. It is then extracted three to four times with ether until supernatant fluid is free from coloring matter. The ether is removed by filtration and discarded. The ground muscle is dried at room temperature until free from ether odor and extracted with

95 % alcohol in the proportion of 5 c.c. to 1 gm. The extraction is carried out 9 days in ice-box and 1 day at room temperature. The alcohol is filtered off and a given quantity cholesterinized by adding 4 mgm. of cholesterin per 1 c.c.

The dilution of the antigens for the tests. — The alcoholic antigen is diluted with salt solution (0.85 % NaCl) in the proportion of 1: 2. The cholesterinized antigen is diluted in the proportion of 1: 3. Pipette 1 c.c. of alcoholic antigen into a 10 c.c. cylinder (inner diameter about 3/8 inch) and from another cylinder add 2 c.c. of salt solution and mix by inverting back and forth. This mixture has no tendency for precipitation at low room temperature. The cholesterinized antigen, on the other hand, will frequently precipitate at this temperature after dilution with salt solution. To overcome this tendency add the salt solution rapidly and immediately mix by inverting back and forth. In the case of this antigen, 3 c.c. of salt solution per c.c. of antigen are employed. This mixture should be kept in the incubator when not in use. Both the alcoholic and cholesterinized antigens after dilution should appear opalescent and absolutely clear.

The test proper. — Measure two 0.3 c.c. quantities of clear inactivated ($\frac{1}{2}$ hour at 56° C.) serum into two small test tubes. Add 0.05 c.c. of the diluted alcoholic mixture to one tube and the same amount of the cholesterinized mixture to the other. Shake vigorously for several minutes. The strongly positive serums will show, in most cases, spontaneous precipitations with the cholesterinized antigen and in a lesser number of cases, with alcoholic antigen. The tests are incubated over night at 37.5° C. and the final results read the next morning.

The micro tests are performed with 0.03 c.c. of serum and 0.005 c.c. of antigen. These should be mixed thoroughly and tubes corked to prevent evaporation during the incubation period.

The controls, both with the regular and micro procedures, should consist of not fewer than 3 positive and 3 negative serums.

The reading of results. — The tubes should not be shaken

again before the first reading. The specific precipitates are suspended in the medium. The + + + + reactions will show either one or several clumps. The + + + reactions will show comparatively large flocculi or granules. The + + reactions will show flocculi or granules of a lesser size, but large enough to be unmistakable. The + and ± reactions are best seen by slanting the tubes and observing the upper point of contact between the fluid and tube wall. When one is working with older serums, there is a tendency for a fine whitish precipitate settling on the bottom of the tube. Shaking re-dissolves these precipitates. It is therefore well to pick out the tubes showing questionable precipitates and shake them vigorously for about a minute and repeat the reading. If the precipitate is so fine that there is a question whether or not it represents a specific precipitate, the test may safely be considered negative.

The results are read in accordance with the following scheme:

1. A precipitate consisting of one or several large clumps = + + + +.
2. A large flocculent precipitate = + + +.
3. Moderate-sized flocculi or granules = + +.
4. Small-sized flocculi or granules = +.
5. Fine flocculi or granules = ±.
6. Negative precipitation = –.

Bureau of Laboratories
Michigan Department of Health
Lansing, Michigan.

THE FILTRATION PLANT FOR DETROIT

THEODORE A. LEISEN

INTRODUCTORY

Before launching into a description of the Filtration Works now under construction for the city of Detroit, it would seem desirable, as a prelude thereto, to touch casually on the history and development of filtration in its application to public water-supplies, to describe the common types of filters, and to give a brief general description of Detroit's sources of supply and the successive stages which led to the adoption of plans for the filter.

HISTORICAL

To ignore all allusion to the more or less vague accounts of the application of certain coagulants for clarifying water-supplies in the early centuries, and to confine the question at issue to the later field of definitely established facts, it may be asserted that the earliest authentic record of filtration as applied to public water-supplies was in 1829 at East Chelsea, England, when, upon recommendation of the Royal Commission on the Metropolitan Water Supply, the Chelsea Water Company constructed and placed in operation a slow sand-filter for the treatment of the water supplied by this company. Twenty years later, 1849, England suffered from a severe cholera epidemic, and the theory was then first advanced that cholera and analogous ailments were water-borne diseases, and the epidemic traced directly to polluted waters. Because of recognition of the soundness of this theory, the London Metropolitan District was compelled, by an act of Parliament, to filter its entire water-supply, thereby stamping filtration with official governmental approval.

The adoption of filtration as the most effective means of safeguarding the urban public against water-borne diseases was very gradual. Many scientists as well as laymen remained skeptical regarding its merits until the cholera epidemic of Hamburg, Germany, in 1892 gave such convincing proof of its efficacy that further controversion was futile. During that year Hamburg, one of the few German cities having unfiltered water, lost over 8,600 persons by death from cholera, while Altona and Wandsbeck, separated from Hamburg only by imaginary boundaries, but having filtered water-supplies, remained comparatively free from the disease. Along one street which forms the boundary between Hamburg and Altona, cholera was rampant on the Hamburg side, whereas the Altona side was free from it, and the only difference was that Altona had a filtered water supply and Hamburg did not. Many later instances and much data, collected both in Europe and in this country, confirmed the evidence furnished by the Hamburg case, but none was more convincing.

FILTRATION IN THE UNITED STATES

The development of filtration in the United States was even less rapid than in Europe. The earliest serious efforts towards investigation, with a view to its adoption in this country, were made in 1866, when Mr. James C. Kirkwood was commissioned by the city of St. Louis to make a study of European filters, and to report on the feasibility of its application for treatment of the Mississippi River water. This exhaustive report, published in 1869, was the first important treatise on the subject issued in the United States, and it may be of interest to note, — as indicative of conservative municipal procrastination, — that fifty years elapsed after the publication of Kirkwood's report before St. Louis installed a filtration plant, although for several years its water-supply was greatly improved by coagulation and sedimentation.

The first municipal filter in the United States was constructed for Poughkeepsie, N. Y., in 1874 to treat water taken from the Hudson River, and this was followed two years later by one at Hudson, N. Y. Both of these were uncovered slow sand filters of crude construction, each with a daily capacity of from two to three million gallons.

During the decade following a limited number of small filters were installed in different sections of the country, and this period gave birth to the so-called mechanical filter, a product of American inventive genius, involving the use of coagulants, convenient cleaning methods, and mechanical agitators. In 1892 the publication of the report of the Massachusetts State Board of Health describing the exhaustive experiments on filtration made at the experimental station at Lawrence, Mass., and the subsequent construction of the slow sand filter in that city, gave a needed impetus to the subject, making the art of filtration a real and recognized issue in this country, and was the inception of a period of rapid and progressive increase in the number of communities furnished with filtered water. In 1890 there were but 300,000 people in the United States supplied with filtered water, while at present more than twenty million inhabitants of our towns and cities are enjoying the benefits of filtered water-supplies, and as a natural sequence, the typhoid death-rate per hundred thousand of urban population has decreased from 48 in 1890 to 13 in 1917.

TYPES OF FILTERS

If the occasional eccentric or freakish types are eliminated, filters may be divided into two classes, slow sand and mechanical or rapid sand filters. The names "slow" and "rapid" designate the principal characteristics of the respective types. The latter, as previously noted, was a development of this country, and offered a solution for the treatment of the highly turbid waters of our western rivers, with which the slow sand filter never could have coped successfully.

In both classes, the water in the process of filtration passes downward by gravity through the sand beds and both classes use sand as a filtering medium with a substratum of coarser material for drainage purposes, and a system of collector drains or pipes at the bottom of the coarse gravel for conveying the filtered effluent through proper conduits to the filtered water reservoir. The slow sand type, as the name implies, is operated at a very slow rate, averaging less than four million gallons per acre per day and the raw water conveyed to the sand beds without any previous treatment, except such as may be afforded by plain subsidence where storage reservoirs are available. Consequently the rate of flow of the water through the sand must be greatly retarded, as otherwise fine particles of mud or other matter held in suspension, including bacteria, will penetrate the sand bed and a certain portion will pass through.

The rapid sand filters are commonly operated at rates of from one hundred million to one hundred and twenty-five million gallons per acre per day, and experiments recently conducted in Detroit have shown that rates up to one hundred and eighty million gallons may be employed without impairing efficiency. One of the typifying factors of the rapid sand filter is the preparatory treatment of the water before it reaches the sand beds. This consists of the use of a coagulating medium, usually aluminum sulphate or sulphate of iron, injected before the water reaches the mixing chamber, where it is decomposed by combination with the alkaline constituents generally present in the water (or artificially supplied when lacking), forming a gelatinous precipitate known as aluminum hydrate or hydroxide of iron, according to the coagulant employed, which has a tendency to unite the minute particles of suspended matter into visible masses and coincidentally enmeshing the bacteria. The water then passes through a coagulation basin at low velocity where the greater portion of the coagulated sediment (fifty to seventy-five per cent) is precipitated. The residue deposited on the surface of the sand bed immediately forms a coagulum scum which, acting as a porous screen, impedes the passage of suspended matter and the enmeshed bacteria, and by the immediate formation of this coagulum surface prevents the passage of all matter held in suspension, even at the rapid filtering rate.

The other important difference between the two types of filters is the method of washing. All filter beds become clogged on the sand surface after passing a given quantity of water, a condition which is governed principally by the turbidity of the raw water. Normally a filter will pass from fifty million to three hundred million gallons per acre of sand before the clogging creates a prohibitive loss of head, and then the beds must be cleaned. Slow sand filters are cleaned by scraping off a thin upper layer of sand which has become impregnated with an accumulation of mud, work that must be done by manual labor with shovels or spades, a crude and tedious operation. Rapid sand filters are washed by reversing the flow of filtered water, which, bubbling up from the bottom, holds the sand practically in suspension, washing off the adhering coagulum and mud which overflows into the wash-water troughs and is conveyed thence to the sewer. The actual washing of a rapid sand filter bed occupies about four minutes.

The relative merits of slow sand and rapid sand filters was a subject of ardent controversial discussion among sanitarians during the earlier stages of filter development in this country, but in the last decade the rapid sand type has forged to the front, and today over seventy-five per cent of the filtered water delivered to American municipalities is furnished through the medium of rapid sand filters, — the same type as adopted for Detroit.

DETROIT'S SOURCE OF SUPPLY

For a number of years prior to 1874, Detroit's pumping station was located at the foot of Orleans Street, about three miles farther down-stream than the site of the present stations, and the supply was taken direct from the river at that point. The rapidly increasing pollution of the river water made the Orleans Street site unsafe as a source of supply, and in 1873, in an effort to get beyond the contaminated zone, the property on East Jefferson Avenue was acquired and construction started on a new pumping station, a small settling basin and an intake pipe. This intake and two subsequent ones all extended but a short distance out into the river, and with the increasing growth of population above this point, within a comparatively few years the supply began to exhibit evidence of excessive pollution, virtually a repetition of the earlier experience at the Orleans Street site. To combat this growing menace and to provide a larger supply, a new intake crib and connecting tunnel were determined on, the crib to be located farther up-stream and sufficiently distant from the shore to avoid normal pollution-laden currents. In 1905 the fulfilment of this project was effected by the construction of the present intake crib and tunnel.

Since that date the water-supply for the city of Detroit has been taken from the upper reach of the Detroit River, close to where Lake St. Clair flows into it, through an intake crib located near the head of Belle Isle and opposite the easterly section of the city, at a distance of 3,200 feet southeastwardly from the shore line at the Pumping Station Grounds, or as locally designated, Water Works Park. A brick tunnel ten feet in diameter extending under the river connects the intake crib with the shore shaft and junction well, whence it distributes through one ten-foot diameter conduit and several conduits of lesser size, either into the small settling basin or directly through other conduits to the suction wells of the pumping stations.

In the efforts to secure a pure source of water-supply through the medium of shifting locations of the intakes, the ultimate goal had been attained in the present intake, as there is no evidence to warrant the belief that a further extension of the tunnel towards Lake St. Clair would be productive of better results. Hence, as a natural sequence, some method of treating the water appealed to reason as the logical solution for safeguarding the supply, so in 1913 calcium hypochloride was resorted to as a sterilizing agent for ameliorating a typhoidal condition which gradually had been exceeding a normal stage. This was followed in 1914 by the use of liquid chlorine, and this treatment has continued uninterruptedly to the present time. That the treatment was necessary is evidenced by the fact that for eight years prior to 1914 the death-rate from typhoid fever averaged twenty-two per hundred thousand of population, while for the eight years from 1914 to 1921, inclusive, the average typhoid death-rate was less then eleven per hundred thousand.

The foregoing statements outline the evolutionary efforts taken to provide a better water-supply, and mark the progressive steps which pointed inevitably — if undesignedly — towards filtration.

After some years of ineffectual attempts to enlist official interest in the question of filtration, authority was finally granted in 1917 to construct a small experimental plant, the chief object of which was to prove conclusively the correctness or falsity of the hypothesis previously advanced, that Detroit River water could be treated at a rate considerably in excess of what heretofore had been deemed feasible or in general use. The experimental plant was completed and put in service the latter part of 1917, and several years' operation demonstrated that the water could be filtered with efficient results up to two hundred million gallons per acre per day, or slightly over three gallons per minute per square foot of effective sand area.

The operation of the experimental filter, and the free distribution of its output, with the attendant publicity derived therefrom, in combination with a growing distaste for chlorine flavor, were potent factors in moulding public opinion in favor of filtration. In August, 1920, therefore, when the citizens of Detroit, by a referendum vote, passed on the question of issuing bonds to the extent of twelve million dollars, one half of which was expressly designed for the filtration work, the verdict was overwhelmingly in its favor, and filtered water for Detroit became an assured prospect.

PLANS FOR FILTER

A general plan for the proposed filtration plant was prepared by the writer in 1916–1917, founded on a careful study of existing works in their relation to the proposed plant, particularly with reference to the utilization of the low lift pumps as boosters pending completion of the filters. In 1919 these plans were reviewed and compared with other suggested schemes, and after careful consideration the original plans were adhered to in practically every detail, except where equipment previously ordered necessitated a change.

The proposed filter for Detroit will have the largest capacity of any filtration plant in the world. The plans in their entirety contemplate a plant having a daily filtering capacity of from 320 million to 360 million gallons, with provisions for meeting peak loads of short duration up to a 400 million gallon rate. No provision has been made for any future extensions to the filter, because, when the average daily consumption approximates 300 million gallons, the limits of capacity of the intake tunnel and the pumping equipment, as well as the filter plant, will have been reached, and any further development on the present site would be injudicious, if not impossible.

The general design of the new filtration plant is divided into three separate structures; first, the Low Lift Pumping Station; second, the Coagulation Basin, Filter Beds and Wash-Water Tanks, all combined under one roof; and third, the Filtered Water Reservoir.

LOW LIFT PUMPING STATION

The Low Lift Pumping Station, 65 feet wide by 175 feet long, is located immediately in line with the ten-foot and six-foot diameter conduits which were torn out for the full length of the station. The water flows from these conduits into a screen chamber at the southerly end of the station where seven electrically-operated revolving screens, each six feet wide by twenty-five feet high from base to normal river level, intercept all coarse floating material and ice formation; with provision for cleaning the screens while in operation. The screening element is composed of copper wire, 14 B. W. gauge, three meshes to the inch, which leaves clear openings about one-quarter inch square, and provides a clear waterway through the screens in the ratio of seven to one of the cross-sectional area of the intake conduit. From the screen chamber the water flows into the suction well twenty-six feet wide by eleven feet deep directly under the floor of the pumping station.

The engine floor which was placed low to provide short suction connections to the pumps, is twenty feet below ground level, with a broad balcony around the interior of the station at entrance level. The pumping equipment consists of four 700 h. p. and one 400 h. p. motor-operated centrifugal pumps, having a combined capacity of 465 million gallons daily, designed to deliver water to the mixing chamber of the coagulation basin against a head of about thirty feet. Arrangements have been made to utilize this station as a booster station to furnish water to the suction pipes of pumping engines in the old station, should a combination of extremely high consumption and low river water make such service necessary prior to completion of the filter plant.

FILTERS

The Filtration Works proper are all under one roof in a structure 480 feet wide by 810 feet long, with a three-story building at the center of the north end which forms the main entrance to the filters, and constitutes the tower for the wash-water tanks, providing also space for laboratory and offices. Each of the two tanks has a capacity of 75,000 gallons, a quantity sufficient to wash one filter bed. Immediately south of the entrance building are the eighty filter beds occupying a space 480 feet by 270 feet, and south of these the coagulation basins, with the mixing chamber and chemical feed and storage space between.

The mixing chamber is considerably smaller in proportion to the capacity of the plant than those which have been provided in most recent filter installations, but is designed to provide thorough and rapid mixing at high velocities, by means of a diversified system of baffles. The chamber is 18 feet wide and 240 feet long, covered for the greater portion of its length so the water may start through under a slight pressure. Alternate sets of vertical and horizontal baffles are placed throughout the full length of the channel, and as the coagulant solution is introduced where the water flows in from the Low Lift Station, there will be ample opportunity for a perfect admixture by the time the coagulation basins are reached. Chemical bins and mixing apparatus and chemical storage space are provided above the mixing chamber; dry feed chemical mixers are used instead of large solution tanks.

The two coagulation basins cover a space 480 feet wide by 525 feet long, and have a total capacity of 30 million gallons, providing a subsidence period of over two hours when the whole filter is operated at its maximum rate. During cleaning periods, which probably will not occur more than once a year, one basin will be closed off while the other remains in operation. Each basin is divided for the greater portion of its length by a baffle wall, so that the coagulated water flows at a low velocity a distance equal to twice the length of the basin. The bottoms are sloped to low points where mud valves permit flushing the accumulated sediment into the sewer.

On account of prevailing low winter temperature, the coagulation basins will be covered, and a steel trussed roof type of construction was adopted as affording a better opportunity for observation of coagulating effects, and also permitting a more pleasing architectural treatment at no additional cost. This arrangement makes the external elevation of both filters and coagulation basins practically the same.

Between the coagulation basin and the filter beds a conduit extends for the full width of the structure, from which the coagulated water flows into each of the conduits feeding a double tier of eight filter beds.

The filter consists of eighty beds divided into five double rows of eight beds with a pipe gallery between each set of double rows; the operating floor forms the cover for the pipe galleries. Each bed has an effective sand area of 1088 sq. ft., or one-fortieth of an acre, and has a filtering capacity of four million gallons daily at a one hundred sixty million rate, or four and one-half million gallons at a one hundred eighty million gallon rate. The filter is covered with a steel-trussed roof, having continuous raised monitors over the operating galleries, thereby affording provision for ample light and ventilation.

Arrangements for sterilization through the medium of liquid

chlorine are provided in a space over one of the filter beds on the westerly side where, if found necessary, chlorine can be introduced at the end of the main filtered-water collector before it enters the filtered-water reservoir.

All water conduits up to the point where the water is delivered on to the filter beds through 24-inch gate-valves are of concrete, and in the design of these the effort has been made to convey the coagulated water to the beds with the least disturbance possible, to prevent the breaking up of the "floc." All conduits for the collection of filtered water and its conveyance to the filtered-water reservoir also are of concrete.

The wash-water troughs are cast iron, thirty-four feet long, extending from the rear end of the filter bed to the wash-water channel at the front or gallery end of the beds, and supported at the centre from concrete T beams, which form walks across the middle of the beds.

The strainer system, which collects the filtrate after it has passed through the sand bed, and also serves as a distributing multiple nozzle for the wash water during washing periods, is composed of a manifold sytem of cast-iron pipe, branching from two main cast-iron collectors built partially into the concrete bottom, the small branches two inches in diameter and the larger or main branches seven inches in diameter. These pipes are tapped along the upper surface and brass strainer cups screwed in at intervals of about six inches over the entire bottom.

The filtrate from each bed will flow from the two main collectors into a connecting pipe which in turn connects with a Venturi-type rate-controller. The water passes through it into the filtered-water conduits under the pipe-gallery floor and on through the main filtered water-collector, which traverses the building from east to west and across the north end of the Low Lift Pumping Station, into the Filtered Water Reservoir. The rate-controller is the governing factor in filtration, as it regulates automatically the quantity of water passing through the filter in accordance with any given rate at which it may be set, and indicates and records the quantity of water filtered.

The filtering medium consists of a level bed of sand twenty-six inches deep, overlying a bed of gravel seventeen inches deep, graduated in size from one and one-half inches at the bottom to one-tenth of an inch at the top. The sand will be relatively coarse, the grains varying from 0.40 to 0.90 millimeter, having a uniformity coefficient of 1.6 to 1.7 and an effective size of 0.40 to 0.50.

Each filter bed has separate control, manipulated from an operating table placed opposite the centre of the bed, and directly above the pipe gallery, upon which are placed the gauges for indicating and recording operations, and the various keys which operate the four controlling hydraulic gate-valves. All valves on the filters and in the coagulation basins are hydraulically operated.

FILTERED WATER RESERVOIR

The Filtered Water Reservoir, which will occupy the site of the existing settling basin, will be an all-concrete structure 360 feet wide by 810 feet long, with flat slab reinforced concrete floor and roof, the latter supported on concrete columns. The roof will be covered with earth and sodded or seeded to form an addition to the park area, to be utilized either for tennis courts or other recreational purposes. The reservoir when full will contain approximately forty million gallons, and will serve as a balancing factor to meet the fluctuating conditions of consumption without disturbing the rate of filtration.

GENERAL CONSTRUCTION

The substructure of the Low Lift Pumping Station, all of the filtration structures up to the water level in the coagulation basins and filter beds, all conduits, the two wash-water tanks and the chemical bins were constructed of reinforced concrete. The superstructure of the chemical and wash-water buildings with the exception of the curtain walls are reinforced concrete also. The superstructure of the pumping station is brick and steel construction, and the superstructure of all sections of the filtration plant is brick with gray pressed-brick and stone trim for all exterior walls, including the veneering of all exposed exterior concrete walls.

The general character of the bottom of coagulation basins and filter beds is beam and slab construction, the beams as well as wall footings having been designed to bear directly on concrete piles. The base area of the whole plant is built directly on the ground, with the pile heads projecting from six to twelve inches above the excavated level of the ground surface, but no dependence was placed on the bearing value of the soil, as the entire weight of structure and contents was calculated to be carried by piling.

On account of the unstable nature of the soil, the pumping station, filters, coagulation basins and filtered water reservoir had to be supported on pile foundations, nearly thirty thousand piles being required for the purpose. Approximately sixty-five hundred all-concrete piles, nine thousand two hundred composite ones of concrete and wood, and fourteen thousand wooden piles will have been driven when the filtered water reservoir foundation is completed. The necessity for the pile foundation adds over a million, seven hundred thousand dollars to the cost of the plant.

CONSTRUCTION METHODS AND PROGRESS

Preparatory to the work of constructing the filter plant, borings were made over the proposed site, and a number of test piles driven, and the information derived therefrom proved conclusively that piling would be required under all portions of the structure. As the greater portion of the filter foundation is above low river level, concrete piles were determined on as the proper type. After competitive bidding, the "Raymond" pile was selected, a type in which a corrugated reinforced tapering steel shell is driven with a steel core or mandril, and after the core is withdrawn the shell is filled with concrete. On account of the fact that subsoil conditions varied greatly, and that the limit of length of the Raymond pile was about thirty-eight feet, composite piles were used where the thirty-eight foot piles did not provide ample bearing value. These consisted of wooden piles of varying lengths, which were driven first and followed by a concrete pile twelve to fifteen feet long, with the wooden section doweled into the concrete portion. The result was a pile having a total length of from sixty to seventy feet, with the wooden portion well below the water-line.

The pile foundation work was completed early in 1921, and the concrete structure of the filtration works started immediately thereafter. The first concrete was poured on June 3, and the work completed on November 15, 1921. This contract comprised 10,000 cubic yards of excavation, 50,000 cubic yards of concrete, 2,200 tons of reinforcing steel, 96 tons of structural steel, and the placing of 770 tons of cast-iron specials.

CONTRACTOR'S PLANT

The equipment of the contractor consisted of two concrete mixers, one locomotive crane, an industrial railway with gasolene locomotives and cars, four five-yard buckets and two cableways. The concrete mixing plant was located at the south end of the grounds near the canal, at which point the gravel and cement were delivered by barges. The gravel which was stored on the ground was handled by the locomotive crane, and dumped into an elevated hopper-bottom loading bin, from which it was chuted by gravity into the charging hoppers of two one-yard concrete mixers. The cement was raised to the mixing floor by a construction elevator. The concrete mixers, set about twenty feet above the ground, delivered the mixed concrete into a five-yard hopper, which in turn dumped it into the five-yard buckets carried on the small flat cars.

The loaded buckets were immediately conveyed to a point under the cableway, at the south end of the work, and by the cableways transferred to any point within the entire area of the filter plant and the concrete deposited in place. Each flat car carried one bucket, a train consisting of two cars, one for the empty, and one for the full bucket.

The cableways were an important feature of the contractor's plant. Each cableway consisted of two timber towers 110 feet high, running on seven parallel lines of rails at each end and counter-balanced to withstand the strain of the cable. The carrying cable extended north and south from head tower to tail tower, a distance of 1,000 feet, along which the buckets were hauled by a lighter auxiliary cable; the lateral or east and west motion was accomplished by moving the connected pair of towers on the tracks. The towers were moved at a rate of 200 feet per minute, and the buckets or other load on the cables 1,200 feet per minute.

Not only the concrete, but all reinforcing steel, structural steel, cast-iron pipe and concrete forms were conveyed by the cableways, and excavated material was removed in the same manner. For a job of the nature of the filter plant or a concrete reservoir, covering a large area, but of no great height above the ground, the cableway would seem to offer the best handling and conveying solution.

COMPLETION OF FILTER PLANT

Bids for the Filtered Water Reservoir, which will cover an area of nearly seven acres, and will require over 32,000 cubic yards of concrete, were received on February 14, and construction will start shortly. The structural steel for the roof of filters and coagulation basin is practically complete, and a contract for the brick superstructure and roofing will be let in March. The filter sand and gravel and all filter equipment will be installed during the present year, and it is expected that the complete plant will be in operation by May, 1923. The total cost of the filtration plant will approximate five million dollars, equivalent to $15,000.00 per million gallons capacity.

Department of Water Supply
Detroit, Michigan

THE HABITAT OF THE BROOK TROUT IN MICHIGAN

T. L. HANKINSON

One purpose of the survey of the fish of Michigan waters, which has been in progress for a number of years, is to find out as much as possible concerning the preferred habitats of the different species of fish found. During several summers prior to 1919, I conducted fish survey work in this State through the Michigan Geological and Biological Survey and under the direction of the chief naturalist of this Survey, Dr. A. G. Ruthven. Many collections were made in many bodies of water including some dozen streams containing brook trout. In the preparation of this paper, the data thus obtained have been used and also data obtained from the examination of brook trout streams in New York State (Hankinson '22) made while I was working for The Roosevelt Wild Life Forest Experiment Station during two summers.

From these field studies, the environmental conditions with most evident importance to the brook trout are as follows: (1) shore vegetation, (2) submerged aquatic plants, (3) topography of stream bed, (4) bottom material, (5) volume of water, (6) nursery streams, (7) substances in water, and (8) associated animals. Each of these sets of conditions constitutes a rather distinct factor influencing the brook trout and is here treated separately.

1. *Shore vegetation.* — An abundance of trees, shrubs or other tall plants bordering a stream is generally considered to be essential to the continued existence of brook trout in it. Evidence of the importance of such a condition is the finding, as I have often done, of the brook trout present only in those parts of extensive stream systems where such plant growths were found. No doubt high vegetation is important in giving shade and low water temperature, both of which brook trout need, and it also

furnishes much food in the nature of foliage insects that frequently fall into the water to become the prey of the trout. Some protection from enemies is secured, even from anglers who find it difficult or impossible to fish from marginal thickets. In fact I have known such thickets to prevent all the good brook trout being taken from a stream by a few individuals in a short time.

Feeding grounds are produced for very young trout by partly submerged plants like sedges, which grow in clumps and hold the soil so as to maintain these shallow areas, allowing algae with accompanying insect food to develop there. Undercutting of banks by currents furnishes places of seclusion for large trout through roots retaining the superficial soil. Pools are important in brook trout streams, and these are sometimes produced through the partial damming by logs, brush or other vegetal débris. Plants along streams prevent extensive soil-wash accompanied by continued turbidity, which is unfavorable for brook trout (Titcomb, '15, p. 9). It is well known that the removal of trees and shrubs along trout streams causes the disappearance of brook trout from them. Such seems to be the cause of the ruin of many trout streams in Michigan. Nevin ('18, p. 48) in writing of conditions in Wisconsin says that in many counties of the State intensive farming has caused the removal of brush and trees from the banks of streams, thus destroying the conditions that provided natural haunts for the brook trout. The restoration of such conditions is often advocated, and Nevin here suggests the planting of willows and tag alders along the stream for this purpose.

2. *Submerged aquatic vegetation.* — The true aquatic plants, those that grow wholly submerged or nearly so, have not been found abundant where there were many brook trout except the young. Little trout two or three inches long are often abundant about thick alga growths in marginal shallows, or among the tufts of waving Cladophora in midstream. In such situations the little trout undoubtedly find much food, which is harbored by the algae. Here are many insects and crustaceans suitable for food for young trout, and — a very important item — midge

larvae (Needham, '03, p. 205) are abundant about algae. No doubt the best planting places for fingerling brook trout are at these quieter parts of the stream where there is conspicuous alga growth, and not in the deep swift waters, where I have seen cans of young brook trout dumped.

3. *Topography of stream bed.* — The bed of the stream should be irregular in inclination, as steps, rather than a steady incline, so there will be quiet pools and intervening stretches of rapid water, both important to the life of the brook trout. The pools are for retirement in warm weather; when large, they may become winter homes. They are also feeding places. Rapid water is important to brook trout, but the fish readily adapt themselves to quiet water as in ponds. One of the best natural brook trout waters found by me in western New York was a series of marsh pools that were slightly connected and apparently stagnant. Here trout fishing was excellent, and the young were numerous and were making long excursions into the dense vegetation for food. An important reason for the abundance of sizable trout here was undoubtedly the failure of anglers to recognize the place as brook trout water. Conditions similar to those found in this spot are very prevalent in Michigan, and further studies of the brook trout in Michigan may reveal the possibility of establishing them in many of our marsh waters.

4. *Bottom material.* — The brook trout waters examined by me had bottoms of diversified character, of muck, marl, clay, sand and gravel. The latter appears to influence brook trout more than the other materials since its accumulations form shallows with riffles. The aëration of the water on these riffles is in all probability beneficial to the trout, especially during long periods without freshets. Under the stones of riffles live many invertebrates, such as larvae of caddice-flies, May-flies, chironomids, snails and other forms, nearly all of which are fed upon by brook trout (See Kendall, '14, p. 79; '18, p. 543). Brook trout feed upon minnows; the larger fish, according to Kendall, may neglect insects for minnows. Gravel shallows about riffles are important breeding places for the common minnows of the trout streams, as well as for other small fish that may furnish

food for trout, such as darters and suckers. Gravel bottom also is used by the brook trout for breeding purposes.

5. *Volume of water.* — It is very probable that the larger the stream the larger the trout it will produce. Kendall ('14, p. 91), in writing of this species notes: "As a rule trout are found to grow in proportion to the size of the stream in which they are found." I have not studied the large trout streams of Michigan extensively, but in the small ones there does not appear to be any close relation between the size of the streams and the size of the brook trout in them. No very large trout have been taken in these creeks by me, none over a foot in length, but some near this size have been found in the smallest creeks inhabited by the species. Kendall notes ('14, p. 83) that sometimes brook trout remain in nursery streams and reach maturity there and pass their whole existence in them.

6. *Nursery streams.* — Good trout streams often have small brooks connected with them. These are commonly very narrow, being but a foot or two wide and nearly concealed by vegetation. In these, young brook trout thrive. Sometimes large trout are found also (Kendall, '14, p. 83). I examined two large streams in New York State in which brook trout were scarce. Formerly they had been good fishing places, but were then fished out. In the little tributary brooks of both of these, small trout were numerous; there were a few large ones also. With a cessation of fishing, the main streams may again become rehabilitated by brook trout from the nursery streams.

7. *Substances in water.* — Much is said and written nowadays about the spoiling of trout streams by sewage, factory and creamery wastes, sawdust and other artificial introductions, but little definite information appears to be available from scientific studies of the water content. Nash ('08, p. 65) gives pollution and overfishing as causes of the disappearance of brook trout from much of southern Ontario. I am familiar with two good brooks in New York, apparently favorable for brook trout, in which efforts have been made to introduce them, but with failure, which was very evidently due to wastes from small creameries.

8. *Associated animals.* — The animals associated with brook trout are chiefly its food or its enemies; the latter are predators or parasites. No careful study of the food of the brook trout has been made in Michigan. Kendall ('14, p. 79) discusses the food in a general way. Juday ('07, p. 167) gives data from stomach examination of brook trout from Colorado. Needham ('03, p. 205) gives results of food studies of twenty-seven examples of brook trout from Bone Pond in the Adirondacks. It can not be doubted, however, that in Michigan the brook trout are highly insectivorous. Large ones probably take many minnows also. The habit of taking floating insects that have fallen in the water is well known, and is made use of in the art of angling for brook trout.

Twenty-four other species of fish were found by me in Michigan streams containing brook trout; of these only eight were present in sufficiently large numbers to be of ecological importance to the brook trout. These are here listed in the order of apparent abundance:

Blace-nose dace, *Rhinichthys atronasus*,
Creek chub, *Semotilus atromaculatus*,
Black-striped minnow, *Richardsonius carletoni*,
Common sucker, *Catostomus commersonii*,
Miller's thumb, *Cottus meridionalis*,
Brook stickleback, *Eucalia inconstans*,
Johnny darter, *Boleosoma nigrum*,
Common shiner, *Notropis cornutus*.

To what extent any one of these listed species is fed upon by brook trout in Michigan streams is not known. Some have been eaten by the trout in other regions. Juday ('07, p. 168) notes common suckers from stomachs of Colorado specimens. Smallwood ('18, p. 332) found a black-striped minnow in a brook trout from an Adirondack lake. Kendall ('13, p. 22) says the black-nose dace furnishes food for the brook trout to some extent, but he notes that this minnow occupies the warmer portions of brooks in summer, where brook trout are not found. In the streams I have examined in Michigan and in western New York the black-nose dace were found quite generally distributed in late summer and were often intimately associated with brook trout. On account of the abundance of this minnow in our trout streams, it is worthy of special investigation to determine the extent to which it is a food fish for trout.

The abundant fish associates of the brook trout stream may affect the trout in an inimical way by competing with it for food. It is important, therefore, to determine the food of the common minnows and other fish in a stream in considering it as a place for planting brook trout. Investigations of this kind should be made for each stream, for a species may have very different food in one body of water from that in another. Kendall ('13, p. 22) finds the black-nose dace subsisting mainly on aquatic larvae of insects and upon small insects that fall upon the surface of the water. In a trout stream in New York State, where brook trout were abundant and where black-nose dace were exceedingly abundant, I found filamentous green algae the principal food in the intestines of seven of eleven of these fish opened.

If the examination of many more of these specimens should reveal this material the chief food of the species in the lake, and if the dace were being eaten by the trout, then they would be very beneficial in the stream through making algae indirectly available to brook trout. My observations seem to indicate that brook trout do not thrive in waters where there are large numbers of common shiners and creek chubs, since both of these species, like the brook trout, are surface feeders. Both make themselves nuisances to anglers through beating trout to flies. The incompatibility of these two species with brook trout is probably due to the similarity of food of all three. Creek chubs and common shiners become most abundant in open, sun-lit parts of creeks, especially where there are deep, clear pools. The clearing of banks then gives favorable conditions for these fish; and this may be another important reason why brook trout do not thrive after the banks are cleared of vegetation.

It is commonly assumed that brook trout have many natural enemies among wild birds and mammals, such as herons, kingfishers, minks, and raccoons, but I have been unable to obtain definite data in scientific literature on the extent of the depredations of these creatures on brook trout. It seems to me that the damage must be slight in waters where there are good hiding places, and where the trout are not overabundant. The brook trout's alertness and quickness in obtaining concealment must leave the less wary suckers and minnows as the common prey of the bird and mammal predators that search the trout stream. On shoals, though, yearling and fingerling trout may be feeding in large numbers, becoming easy prey to predators. Herons in such places may do considerable damage to trout, but further information is needed as to the extent of this damage, and that should be obtained through careful field studies. It is generally believed that bass, pickerel, and other piscivorous fish should be kept from trout streams, and certainly they should not be introduced where brook trout are desired. In creeks where brook trout spawn small fish devour their eggs. All of the common brook minnows may be guilty of this offense (Embody, '15, p. 237). It is frequently mentioned in literature that the common miller's thumbs destroy trout eggs, but I have been unable to find published the data on which the statement is based.

Some organisms are injurious to brook trout as parasites, but brook trout are not likely to be injured seriously in this way if they are not too abundant and crowded into a body of water. This is the opinion of Fasten ('12, p. 17), who has given much attention to the parasites and diseases of brook trout. I once found a little stream in Houghton County, Michigan, where brook trout were extremely numerous. Some captured were badly infested by a crustacean parasite, Lernaeopoda.

Under the topic, enemies, there might properly be treated the destructiveness of man to brook trout through overfishing streams, pollution of trout waters, removal of shore vegetation, allowing streams to be roiled and contaminated by domestic animals, especially hogs. In other ways, too, he has been a serious enemy to this valuable species of fish. In a measure this destructiveness has been compensated for through fish culture and by planting.

The observations that I have made on some thirty brook trout streams in Michigan and in New York State have made it very clear to me that results of value, in a purely scientific way and also in an economic way, may be obtained through further and more detailed studies of these streams, with particular attention given to the environment of the brook trout. In an economic way ecological studies must prove of value through our learning more of the proper conditions for a trout stream and of ways of maintaining these conditions; in this way trout fishing will be improved. Through such studies better legislation can be recommended, not only for brook trout waters generally, but for particular waters; for example, in some stream where brook trout may be very abundant but too small for interesting fishing.

It may prove advisable to permit the taking of smaller trout here than elsewhere. Kendall and Goldsborough ('08, p. 53) note such an instance in New Hampshire. There should result, also, better planting methods, and more should be learned as to the kinds of places where plantings should be made in streams. Particular attention should be given to waters in which trout have been introduced successfully. It is unfortunate that we do not know more of the Au Sable River at the time when large brook trout were so abundant there (Clark, '00, p. 81). Conditions in places where plantings have not met with success should also be investigated. In fact each trout stream presents a special problem, and measures for the control of the trout in the stream need to be determined through scientific investigations. A superficial examination during a few hours is not sufficient for any stream, but the studies made of it should be prolonged and intensive, and, as far as possible, they should extend over all the seasons of the year.

State Normal College
Ypsilanti, Michigan

LITERATURE CITED

CLARK, FRANK N. 1900. The Brook Trout and the Lake Trout, Manual of Fish Culture. Bur. of Fisheries, pp. 80–90.

EMBODY, George C. 1915. The Farm Fishpond. The Cornell Reading Courses, IV. 213–252.

FASTEN, NATHAN. 1912. The Brook Trout Disease at Wild Rose and Other Hatcheries. Bien. Report Comm. Fisheries, Wisc., pp. 12–22.

HANKINSON, T. L. 1922. Fish Survey. The Grouse, Buffalo, N. Y. 1: 3, 8, 15, 16.

JUDAY, CHANCEY. 1907. A Study of Twin Lakes, Colorado, with Special Reference to the Food of the Trouts. Bulletin Bur. Fisheries, 26: 151–175.

KENDALL, WM. CONVERSE. 1913. Fishes and Fishing in Sunapee Lake. Bur. Fisheries Doc., No. 783, pp. 1–92.
1914. The Fishes of New England, The Salmon Family. Part I.— The Trouts or Charrs. Mem. Bost. Soc. Nat. Hist., 8: 1–103.
1918. The Rangeley Lakes, Maine. Bulletin Bur. Fisheries, 35: 487–540.

KENDALL and GOLDSBOROUGH. 1908. Fishes of the Connecticut Lakes and Neighboring Waters. Bur. Fisheries Doc., No. 633, pp. 1–77.

NASH, C. W. 1908. Vertebrates of Ontario. Dept. Education, Toronto, pp. 1–107.

NEEDHAM, J. G. 1903. Food of the Brook Trout in Bone Pond. N. Y. State Museum Bulletin, No. 68, pp. 205–206.

NEVIN, JAMES. 1918. Report of Division of Fisheries. Bien. Report State Conser. Comm., Wisc., 1918.

SMALLWOOD, WM. M. 1918. An Examination of the Policy of Restocking the Inland Waters with Fish. Am. Naturalist, 52: 322–352.

SEASONAL VARIATION IN THE NUMBER OF VERTEBRAE OF FISHES

CARL L. HUBBS

I

This study is one of a series dealing with the relation of the environment to the characters of fishes. It supplements a paper in *The American Naturalist*, 56 : 360–372, entitled *Variations in the Number of Vertebrae and other Meristic Characters of Fishes Correlated with the Temperature of the Water during Development.* In that article significant average differences in the number of vertebrae, fin-rays and scales were shown to characterize successive year-groups of two species of fresh-water fishes, *Notropis atherinoides* and *Lepomis incisor*. In the case of all characters studied, in both species, the average number of segments was found higher in the year class which developed at the lower temperature. It was found further, for *Notropis atherinoides*, that significant differences in the average number of vertebrae were developed even within one year class, a drop in temperature during the breeding season entailing then an increase in the number of segments. This second point is the one here followed out.

The material for the present report was all obtained during the summer of 1921, at the University of Michigan Biological Station, Douglas Lake, Michigan. It comprises the young of two species, *Notropis hudsonius* and *Notropis blennius*, and was all secured on the sandy shoal of South Fishtail Bay, Douglas Lake. This shoal was under close observation during the entire summer. The two species of minnows mentioned, as well as the others of the same lake, were readily identified at all sizes, even in the field. The efficient assistance rendered by Mr. Charles W. Creaser in the field work is here gratefully acknowledged.

II

Fortunately, there are available for the present analysis detailed data on the temperature of the water in which the fishes under discussion developed. Temperature readings were taken, usually twice daily, with a calibrated maximum-minimum thermometer. This was kept throughout the summer at a depth of nearly one meter, over deep water, just off the sandy shoal on which the fishes were seined. Since it is almost certainly indicated, by circumstantial evidence, that both species of minnows bred in the vegetation at the edge of this shoal, the temperatures obtained may be regarded as substantially identical with the developmental temperatures involved. These maximum and minimum temperature readings for June, July, and the first three weeks of August, are shown in the accompanying graph.

The exact time of development of the two species was not observed, but can be approximated with sufficient nearness, by reason of the fact that, soon after the hatching stage, i.e., as soon as the yolk-sac is absorbed, the young move in from the vegetation to occupy their characteristic habitat — the extreme shoreward margin of the water. As the development of these fishes is relatively rapid, the time when the number of segments became fixed can be placed a few days prior to the date of their appearance on the shoal.

III

The first young of *Notropis hudsonius*, four in number, 6.5 to 8.0 mm. long, were obtained on June 10. No more were observed until June 14, when they first appeared in abundance; on that date about fifty newly hatched young, 7.0 to 9.0 mm. long, were caught. On June 15 the young fringed the shoreline; a large series, 5.0 to 11.0 mm. long, was secured. The number of specimens of each millimeter group for this and for four subsequent dates is given in Table I.

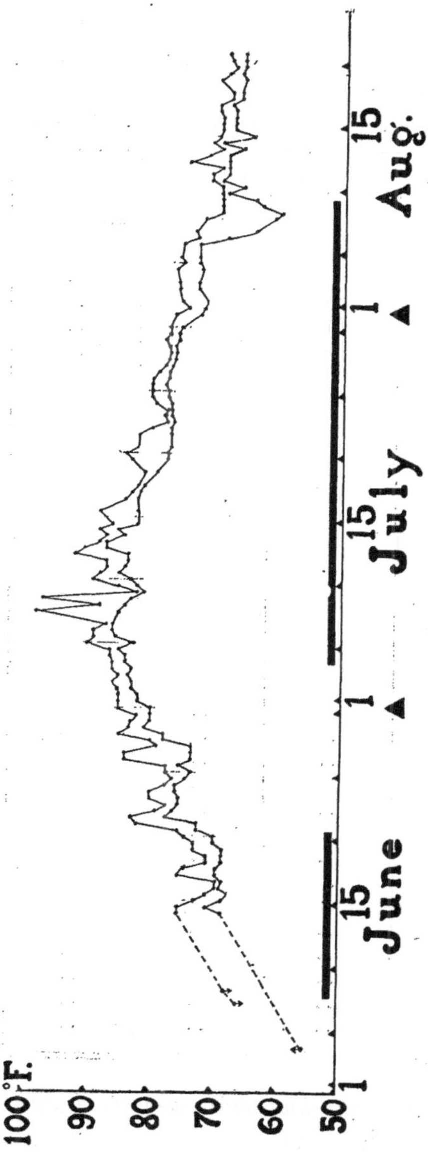

FIG. 7. GRAPH SHOWING WATER TEMPERATURES OF DOUGLAS LAKE IN 1921

The graph shows the maximum and minimum temperatures near the surface in South Fishtail Bay, Douglas Lake, Michigan, during the summer months of 1921. The shorter heavy line indicates the breeding season of *Notropis hudsonius*; the longer one, that of *Notropis blennius*.

FREQUENCY TABLE I

SIZE OF NOTROPIS HUDSONIUS ON DIFFERENT DATES

Length to caudal fin in millimeters	Number of specimens of each size taken on				
	June 15	June 19	June 28	July 7	July 16
5	3	..	..	..	..
6	3	7	..	..	..
7	9	26	..	..	..
8	53	24	..	..	..
9	36	74	..	..	..
10	15	99	2	..	..
11	1	41	5	2	..
12	..	10	5	8	..
13	..	3	11	8	..
14	..	..	32	14	..
15	..	..	46	14	1
16	..	..	48	17	2
17	..	..	38	43	8
18	..	..	25	46	12
19	..	..	10	52	15
20	..	..	7	53	20
21	..	..	1	46	17
22	..	..	1	43	19
23	..	..	..	39	22
24	..	..	..	38	26
25	..	..	..	18	30
26	..	..	..	6	40
27	..	..	..	4	44
28	..	..	..	1	75
29	..	..	..	2	69
30	..	..	..	2	113
31	..	..	..	..	78
32	..	..	..	..	66
33	..	..	..	..	33
34	..	..	..	..	22
35	..	..	..	..	15
36	..	..	..	..	4
37	..	..	..	..	2

These data for *Notropis hudsonius* seem to indicate a rather uniform rate of growth for the young (about 0.7 mm. increase per day); a considerable increase in range of size after the close of the breeding season, and finally, a rather brief period of hatching, commencing a few days before June 10, when the water was rather cold, continuing with increasing intensity during a few following days of rising temperature, and throughout the week (June 13 to 20) of rather uniform temperature.

The series obtained on July 16, the last set listed in the tabulation given above, were alone utilized, for at the size then attained an accurate vertebral count could readily be made of all the specimens. In a rough way, the order of size in this series is inverse to the order of hatching. The largest specimens therefore developed at the coldest temperature, the smallest ones

FREQUENCY TABLE II

Number of Caudal Vertebrae of Each Millimeter Group of Notropis Hudsonius Collected July 16, 1921

Millimeter group	Number of caudal vertebrae				Average number	Probable error
	16	17	18	19		
15	......	......	1	......	18.00	....
17	......	3	7	1	17.82	0.12
18	......	2	5	......	17.71	0.12
19	......	6	9	......	17.60	0.09
20	......	3	12	2	17.94	0.08
21	......	3	4	1	17.75	0.16
22	1	1	12	......	17.79	0.10
23	......	1	14	......	17.93	0.03
24	......	4	20	1	17.88	0.06
25	......	7	19	2	17.82	0.07
26	1	6	25	1	17.79	0.06
27	......	8	35	1	17.88	0.04
28	......	8	18	2	17.79	0.07
29	1	5	26	6	17.97	0.07
30	......	8	24	4	17.89	0.06
31	......	5	26	7	18.05	0.06
32	......	3	12	3	18.00	0.06
33	......	1	11	2	18.07	0.07
34	......	2	6	3	18.09	0.07
36	......	......	......	1	19.00	....

at a warmer temperature; increase in size in this lot corresponds roughly with a decrease in developmental temperature. The relation is not very sharp, however, owing to individual differences in rate of growth following a rather brief breeding season, and especially to the fact that a large proportion of the specimens were hatched out during a period of nearly constant temperature.

The only character counted in this series of *Notropis hudsonius* was the number of vertebrae. As previous work has indicated that it is the caudal rather than the precaudal (abdominal) vertebrae that are subject to modification in number, and as the complex specialization of the anteriormost vertebrae to form the Weberian apparatus renders the count of precaudal vertebrae difficult and subject to error in young minnows, only the caudal or urosome vertebrae were enumerated. The results of the counts are given in Table II. Four hundred specimens were used.

For any one of the millimeter groups, the number of specimens is small and the probable error of the average number of vertebrae is large. Hence it is difficult to trace any tendency toward variation in the number of vertebrae with the size which may exist. The counts may, however, be grouped into three equal divisions, small, medium and large, as is done in the following tabulation:

Millimeter groups	Number of caudal vertebrae				Average number	Probable error
	16	17	18	19		
15–21	......	17	38	4	17.78	0.05
22–28	2	33	143	7	17.84	0.02
29–36	1	24	105	26	18.00	0.03

The largest specimens, which hatched out under the coldest temperatures, thus show the highest average number of caudal vertebrae; the specimens of the intermediate and smallest groups, which hatched out during a warmer period, present a lower average number of vertebrae. Furthermore, there is indicated no definite tendency toward variation in the number of segments, with difference in size of fish, within the two smaller (and younger) groups; the temperature remained nearly constant during the period of development of the specimens comprising these groups.

IV

FREQUENCY TABLE III

NUMBER OF CAUDAL VERTEBRAE OF EACH MILLIMETER GROUP OF NOTROPIS BLENNIUS COLLECTED AUGUST 21, 1921

Millimeter group	Number of caudal vertebrae					Average number	Probable error
	16	17	18	19	20		
18	.....	1	38	30	....	18.42	0.04
19	.....	2	33	24	1	18.40	0.05
20	.....	3	27	19	2	18.39	0.06
21	1	2	14	11	1	18.31	0.10
22	.....	1	12	26	1	18.675	0.06
23	.....	2	17	25	3	18.62	0.08
24	.....	1	34	24	3	18.47	0.05
25	.....	3	44	36	2	18.44	0.04
26	.....	6	44	36	2	18.39	0.05
27	.....	11	54	35	2	18.27	0.045
28	.....	7	52	24	....	18.20	0.04
29	.....	8	37	12	1	18.10	0.06
30	.....	4	30	8	2	18.18	0.07
31	.....	6	19	15	....	18.225	0.07
32	.....	1	14	10	....	18.36	0.08
33	.....	2	16	6	....	18.17	0.08
34	.....	2	5	6	....	18.31	0.13
35	.....	....	5	5	....	18.50	0.11
36	.....	....	2	.....	....	18.00	....
38	.....	....	.....	1	....	19.00	....
39	.....	1	.....	.....	....	17.00	....
18 to 26	1	21	263	231	15	18.45	0.02
Percentages	.....	*4*	*50*	*44*	*3*	....	....
27 to 39	.....	42	234	122	5	18.22	0.02
Percentages	.....	*10*	*58*	*30*	*1*	....	....

The young of *Notropis blennius* were first located on the shoal on July 7, some time after the close of the breeding season of *Notropis hudsonius;* on that date 101 specimens were taken, varying in length to caudal fin from 10 to 17 mm., 14 mm. being the modal size. Throughout the remainder of this month and on into August, the young continued to hatch out and make their appearance on the shoal; each lot augmented in numbers those already present. By reference to the temperature graph it will be seen that this species developed not only in the very hot water of the first half of July, but also in the decidedly cooler (though yet rather warm) water of the latter part of July and the first part of August.

The lot of *Notropis blennius* counted was obtained on August 21, after the end of the breeding season. More than one thousand specimens were taken; they varied in length from 11 to 39 mm., and clustered about two well-defined modes, one at 18 mm., the other at 27 mm. Those smaller than 18 mm. could not be satisfactorily counted; the data are limited to 934 specimens 18 to 39 mm. long. Of these the larger developed at a warmer temperature than did the smaller ones, and, as Table III shows, they present a significantly lower average number of caudal vertebrae than do the smaller specimens.

V CONCLUSION

In both *Notropis hudsonius* and *Notropis blennius* the specimens of a single year class obtained at one restricted locality show significant differences in the number of caudal vertebrae. In both cases those individuals developed under the higher temperatures possess on the average fewer segments than do those hatched out in cooler water. These results accord with others obtained by the writer. An interpretation of their significance will be attempted elsewhere.

UNIVERSITY OF MICHIGAN

ORTHOPTERA FROM GRATIOT COUNTY, MICHIGAN

H. M. MacCURDY

The value of authentic locality lists in the study of problems of distribution is fundamental and is the reason for the appended list of the Orthoptera collected in this locality and adjacent areas. Unless otherwise stated, the specimens are all from this immediate vicinity. It is believed that the list fairly represents the nature of the orthopteran fauna of this part of the State, although it is difficult to estimate or convey a true picture of the relative abundance of the species or their relative economic importance. The latter problem is not undertaken here.

Up to the present time there have been published three local lists and a key of the Orthoptera of Michigan. The lists are as follows: A. P. Morse, *Report on the Isle Royale Orthoptera of the 1905 Expedition;* [1] A. F. Shull, *Thysanoptera and Orthoptera,*[2] from the sand-dune region on the south shore of Saginaw Bay; A. E. Woodward, *The Orthoptera Collected at Douglas Lake, Michigan, in the Summer of 1910;* [3] The key is Special Bulletin No. 83, of the Michigan Agricultural Experiment Station, East Lansing.[4]

It is seen that the three local lists [5] are from rather widely

[1] Pages 299–303 of *An Ecological Survey of Isle Royale, Lake Superior, Michigan* (prepared under the direction of C. C. Adams and published as a part of the Report of the Board of Geological Survey for 1908).

[2] Pages 177–231 of *A Biological Survey of the Sand Dune Region on the South Shore of Saginaw Bay, Michigan* (prepared under the direction of A. G. Ruthven and published as a part of the annual of the Board of Geological and Biological Survey for 1910).

[3] *Michigan Academy of Science, 13th Annual Report,* pp. 146–167. 1911.

[4] R. H. Pettit and Eugenia McDaniel, *Key to Michigan Orthoptera with Annotations.*

[5] Since the reading of this paper another local list of Orthoptera for this State has been published by Theodore H. Hubbell, *The Dermaptera and Orthoptera of Berrien County, Michigan, Occasional Papers of the Museum of Zoölogy, University of Michigan, No. 116.*

separated areas and near the lakes. A list from the central part of the State is important for comparison. It must be remembered, however, that such lists are representative rather than complete; but they are the best means at hand for giving a knowledge of the distribution and nature of the fauna.

Undoubtedly the physiographic features and the vegetation are the chief factors in determining the local distribution of the various species of Orthoptera, though it is rather difficult to show this to be true. Shull found no close correlation between species and habitat. The collecting done in this region shows only a general correlation between species and habitats strongly defined. It does not seem justified, however, to carry this correlation beyond the limits of general application.

The most characteristic habitats within the area are the old lake bottoms and beaches, the marginal moraines, the outwash plains and the marginal drainage basins lying between the different moraines. The moraines and the marginal drainage basins between them extend in a north-south direction across the area, so that a succession of habitats is encountered as one crosses the area in an east and west line, from old lake bottoms on the east to the successive moraines on the west. As yet the materials collected have not been sufficiently analyzed and tabulated to show any close correlation with habitat.

In relation to the question of life zones, one is limited again by the fact that many species are poor students of geography. The list reveals, however, the fact that a few of the species are about on the southern limits of their range within the State, a few near the northern edge, while most of the forms have a wide distribution. If one takes the numbers of families, genera and species represented as an index, and grants that some must have escaped the collector, it is seen that this important order of insects is well represented. It would seem to be justifiable to characterize the area as one on which overlapping in distribution of species occurs. The question of intermediate forms in such an area assumes importance. That does not lie within the limits of this discussion.

The numbers of families, subfamilies, genera and species

represented for the localities listed within the State where collections of Orthoptera have been made show an increase for the more southern localities, illustrating an important principle in distribution especially for these forms. This is shown in Table I. There is a greater number of families, genera and species in the more southern localities.

LOCALITY LIST OF ORTHOPTERA FROM MICHIGAN

Locality	Families and subfamilies	Genera	Species
Isle Royale, Porcupine Mts......	4	11	21
Douglas Lake..................	8	17	30
Saginaw Bay..................	11	27	52
Gratiot County................	14	29	55

For the identification of all but a few of the species, I am indebted to T. H. Hubbell of the Museum of the University of Michigan. For those which were not identified by him, the writer assumes responsibility.

The following list of species represents those that have been taken. Undoubtedly a number of others may be added, which may reasonably be expected to occur here. Such are not included. Detailed notes are also omitted since in general they agree with those of other collectors.

Family BLATTIDAE

Parcoblatta pennsylvanica DeGeer.
Parcoblatta uhleriana Saussure.
Parcoblatta sp. immature.
Blatta germanica L.

Family PHASMIDAE

Diapheromera femorata Say.

Family TETTRIGIDAE

Subfamily TETRIGINEAE

Nomotettix cristatus Scudder.
Acridium granulatum Kirby.
Acridium ornatum Say.
Acridium arenosum angustum Hancock.
Acridium sp. immature.

Subfamily Batrachidinae

Tettigidea lateralis parvipennis Harris.

Family ACRIDIIDAE

Subfamily Tryxalinae

Chloaltis conspersa Harris.
Orphulella palidna Burmeister.
Chorthippus curtipennis Harris.

Subfamily Oedipodinae

Arphia pseudonietana Thomas.
Arphia sulphurea Fabricius.
Chortophaga viridifasciata DeGeer.
Encoptolophus sordidus Burmeister.
Cammula pellucida Scudder.
Dissosteira carolina L.
Spharagemon bolli Scudder.
Spharagemon collare wyomingianum Thomas.
Scirtetica marmorata marmorata Harris.
Circotettix veruculatus Kirby.

Subfamily Locustinae

Melanoplus mexicanus atlantis Riley.
Melanoplus femur-rubrum femur-rubrum DeGeer.
Melanoplus angustipennis Dodge.
Melanoplus bivitatus Say.

Family TETTIGONIIDAE

Subfamily Phaneropterinae

Scudderia texensis Saussure-Pictet.
Scudderia pistillata Brunner.
Scudderia curvicauda curvicauda DeGeer.
Scudderia furcata furcata Brunner.
Amblycorypha oblongifolia DeGeer.

Subfamily Pseudophyllinae

Petrophylla camellifolia camellifolia Fabricius.

Subfamily Conocephalinae

Orchelimum vulgare Harris.
Conocephalus fasciatus fasciatus. DeGeer.
Conocephalus brevipennis Scudder.
Conocephalus nemoralis Scudder.
Conocephalus nigropleurus Scudder.

Subfamily Raphidophorinae

Ceuthophilus nigricans Scudder.
Ceuthophilus brevipes Scudder.
Ceuthophilus terestris Scudder.
Ceuthophilus maculatus Say.

Family GRYLLIDAE

Subfamily Gryllotalpinae

Gryllotalpa hexadactyla Perty.
Tridactylus apicalis Say.

Subfamily Gryllinae

Nemobius fasciatus fasciatus DeGeer.
Nemobius carolinus Scudder.
Gryllus asimilis Fabricius. Specimens agree with *G. scudderianus* Saussure and *G. luctuosus* Serville.

Subfamily Oecanthinae

Oecanthus niveus DeGeer.
Oecanthus angustipennis Fitch.
Oecanthus nigricornus F. Walker.
Oecanthus nigricornus quadripunctatus Beutenmüeller.

The species listed below were taken by the writer at Lake George, Clare County, Michigan, July 9–29, 1921. They are added here for the locality records. They were all identified by T. H. Hubbell.

Parcoblatta sp. juv.
Chorthippus curtipennis Harris.
Orphulella speciosa Scudder.
Arphia pseudonietana Thomas.
Spharagemon collare wyomingianum Thomas.
Spharagemon bolli Scudder.
Circotettix verruculatus Kirby.
Scirtetica marmorata marmorata Harris.
Camnula pelucida Scudder.
Disosteira carolina L.
Melanoplus mexicanus atlantis Riley.
Melanoplus bivitatus Say.
Melanoplus dawsoni Scudder.
Melanoplus stonei. Rhen.
Scudderia curvicauda curvicauda DeGeer.
Scudderia pistillata Brunner.
Conocephalus fasciatus fasciatus DeGeer.

Conocephalus brevipennis Scudder.
Ceuthophilus maculatus Say. (?)
Gryllus asimilis Fabricius.
Oecanthus quadripunctus Beutenmuller.

Among these *Melanoplus stonei* (Rhen) and *M. dawsoni* (Scudder) are noteworthy, while *Circotettix verruculatus* (Kirby) and *Scirtettica m. marmorata* (Harris) are near the southernmost limits of their known range in the State.

Alma College
Alma, Michigan

NOTES ON THE NESTING HABITS OF COTTUS

BERTRAM G. SMITH

One of the most abundant and widely distributed of Michigan fishes is the "miller's thumb," Cottus. Since it occurs in every brook, it invariably attracts the attention of even the amateur zoölogist, yet its habits are very incompletely known. The writer had hoped to make a real contribution to our knowledge of this form, but since circumstances have made it impossible for him to give the subject the attention its interest demands, the following fragmentary observations are submitted in the hope that some one may be stimulated to further work on the problem.

The habitat which I have found most favorable for the study of Cottus is a small brook with stony and gravelly bottom (Fig. 3). During several successive years I have found eggs in various stages of development during April and May, which points to a rather extended breeding season, but my data do not determine its precise duration.

In this brook, the eggs are laid in masses attached to the under surfaces of stones, usually in rather swift water. Under the circumstances, the spawning behavior is difficult to study and has not been observed. The egg-masses are round or oval in outline, flattened and sometimes irregular in contour. The smaller egg-masses (Fig. 1) measure from 20 to 30 millimeters in diameter; each mass consists of about 200 eggs alike in size, color and stage of development. In different masses, the individual eggs vary in size from one to three millimeters in diameter. This difference in size is evidently due in part to parentage, in part to the degree of inflation of the egg envelope. Eggs of different masses vary in color from a creamy white to orange-yellow; this difference cannot be accounted for entirely as a consequence of developmental changes, but is in part inherent in the yolk.

The larger and more irregular egg-masses are usually compound in the sense that each consists of two or three clusters differing in size, color and sometimes in the stage of development of the eggs. Thus the egg-mass represented in Fig. 2 consists of three disc-shaped masses, of which two overlap the one in the center. It seems probable that the "nest" is visited by several females, which spawn successively with the same male.

There is a paternal brooding habit. In almost every case where eggs were found, an adult Cottus was found in the nest; in the other cases it is possible that a fish was present, but escaped observation. Four of these fishes taken from nests were dissected and the sex determined; all were males.

The species, *Cottus meridionalis* Girard, was identified by reference to a recent paper by Hubbs.[1]

Department of Anatomy
University and Bellevue Hospital Medical College
New York City

[1] Carl L. Hubbs. 1919. *Nomenclatural Notes on the Cottoid Fishes of Michigan. Occasional Papers of the Museum of Zoölogy, University of Michigan, No. 65.*

EXPLANATION OF PLATE XI

Fig. 1. Photograph of living eggs of Cottus, attached to a stone. A little more than two-thirds natural size, linear reduction.

Fig. 2. Photograph of compound mass of living eggs of Cottus, attached to a stone. A little less than one-half natural size, linear reduction.

Fig. 3. Nesting habitat of Cottus, at "Riverbrink," near Ypsilanti, Michigan.

Fig. 4. Male Cottus taken from nest, photographed in living condition without use of anaesthetic. The indistinctness of the outline of the head is due to the respiratory movements. Two-thirds natural size, linear reduction.

Fig. 5. Male Cottus taken from nest, photographed after preservation in formalin. About three-fifths natural size, linear reduction.

PLATE XI

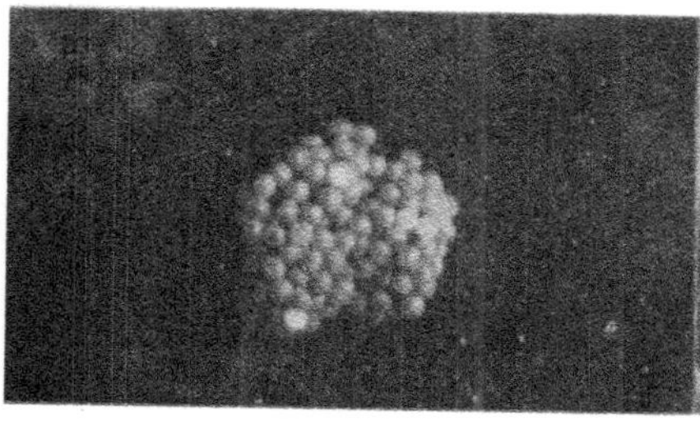

Fig. 1

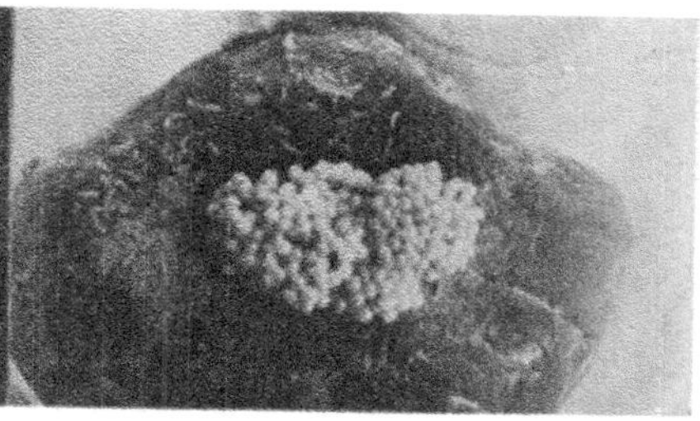

Fig. 2

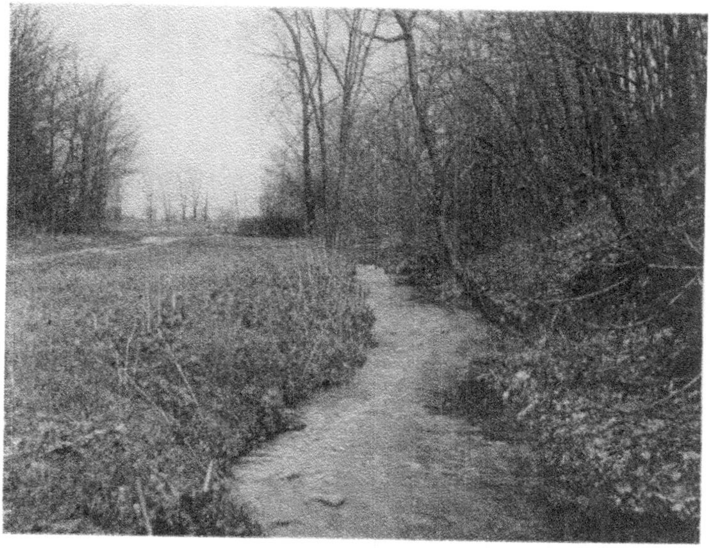

Fig. 3

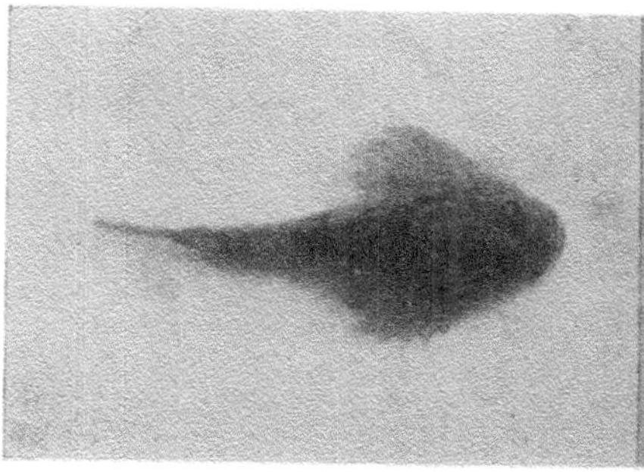

Fig. 4

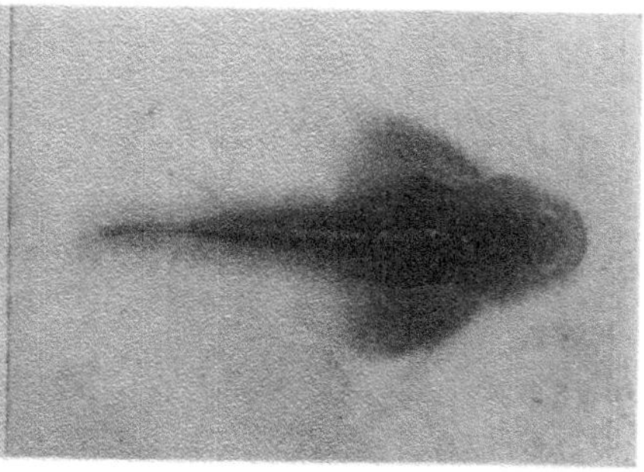

Fig. 5

Nesting Habits of Cottus

INDEX OF AUTHORS AND SUBJECTS

PAGE

BARTLETT, H. H. and LA RUE, CARL D. A Leaf-Fall Disease of *Hevea brasiliensis* Muell.-Arg. due to *Gloeosporium alborubrum* Petch.. 73

BARTLETT, H. H. and LA RUE, CARL D. Diplodia Disease of *Hevea brasiliensis*.......... 91

BAXTER, DOW VAWTER. The Genus Ribes with reference to the White Pine Blister Rust in the West.......... 1

BOWMAN, H. H. M. The Distribution and Pollination of Certain Sea-Grasses.......... 3

BUZZELL, R. S. Some Aspects of the Chlorination Treatment at the Flint Filtration Plant.......... 173

Certain Problems of Labor and the Public Attitude. HERBERT F. TAGGART.......... 143

Diplodia Disease of *Hevea brasiliensis*. CARL D. LA RUE and H. H. BARTLETT.......... 91

Distribution and Pollination of Certain Sea-Grasses. H. H. M. BOWMAN 3

EHLERS, G. M. and LEIGHLY, J. B. *Lipsanocystis traversensis*, A New Cystid from the Devonian of Michigan.......... 155

Extensions of Range of Plants in the Douglas Lake Region, Cheboygan County, Michigan. FRANK C. GATES.......... 47

FARWELL, OLIVER ATKINS. Notes on the Michigan Flora: Part V... 11

Filtration Plant for Detriot. THEODORE A. LEISEN.......... 181

Foreign Investment and Foreign Trade. C. E. GRIFFIN.......... 119

GATES, FRANK C. Extensions of Range of Plants in the Douglas Lake Region, Cheboygan County, Michigan.......... 47

GATES, FRANK C. Meteorological Data, Douglas Lake, Michigan.... 162

GATES, FRANK C. Meteorological Summary, Douglas Lake, Michigan 161

Genus Armillaria in the United States and its Relationships. C. H. KAUFFMAN.......... 53

Genus Ribes with reference to the White Pine Blister Rust in the West. DOW VAWTER BAXTER.......... 1

GRIFFIN, C. E. Foreign Investment and Foreign Trade.......... 119

GUSTAFSON, F. G. Studies on the Hydrogen Ion Concentration of Plant Juices.......... 49

Habitat of the Brook Trout in Michigan. T. L. HANKINSON....... 197

HANKINSON, T. L. The Habitat of the Brook Trout in Michigan ... 197

HARVEY, NATHAN A. Illusions of Orientation.......... 165

HUBBS, CARL L. Seasonal Variation in the Number of Vertebrae of Fishes.......... 207

PAGE

Illusions of Orientation. NATHAN A. HARVEY 165
KAHN, R. L. A Simple Quantitative Precipitation Reaction for Syphilis 177
KAUFFMAN, C. H. The Genus Armillaria in the United States and its Relationships 53
LA RUE, CARL D. Two Unreported Parasites of *Hevea brasiliensis* 69
LA RUE, CARL D. and BARTLETT, H. H. A Leaf-Fall Disease of *Hevea brasiliensis* Muell.-Arg. due to *Gloeosporium alborubrum* Petch 73
LA RUE, CARL D. and BARTLETT, H. H. Diplodia Disease of *Hevea brasiliensis* 91
Leaf-Fall Disease of *Hevea brasiliensis* Muell.-Arg. due to *Gloeosporium alborubrum* Petch. CARL D. LA RUE and H. H. BARTLETT 73
LEIGHLY, J. B. and EHLERS, G. M. *Lipsanocystis traversensis,* A New Cystid from the Devonian of Michigan 155
LEISEN, THEODORE A. The Filtration Plant for Detroit 181
Lipsanocystis traversensis, A New Cystid from the Devonian of Michigan. G. M. EHLERS and J. B. LEIGHLY 155
MACCURDY, H. M. Orthoptera from Gratiot County, Michigan 215
Meteorological Data, Douglas Lake, Michigan. FRANK C. GATES 162
Meteorological Summary, Douglas Lake, Michigan. FRANK C. GATES 161
Notes on the Michigan Flora: Part V. OLIVER ATKINS FARWELL 11
Notes on the Nesting Habits of Cottus. BERTRAM G. SMITH 221
Orthoptera from Gratiot County, Michigan. H. M. MACCURDY 215
Revenue Act of 1921. F. E. ROSS 137
ROSS, F. E. The Revenue Act of 1921 137
Seasonal Variation in the Number of Vertebrae of Fishes. CARL L. HUBBS 207
Simple Quantitative Precipitation Reaction for Syphilis. R. L. KAHN 177
SMITH, BERTRAM G. Notes on the Nesting Habits of Cottus 221
SMITH, HUGH B. Stomatal Behavior of Plants in the Greenhouse in Winter 109
Some Aspects of the Chlorination Treatment at the Flint Filtration Plant. R. S. BUZZELL 173
Stomatal Behavior of Plants in the Greenhouse in Winter. HUGH B. SMITH 109
Studies on the Hydrogen Ion Concentration of Plant Juices. F. G. GUSTAFSON 49
TAGGART, HERBERT F. Certain Problems of Labor and the Public Attitude 143
Two Unreported Parasites of *Hevea brasiliensis*. CARL D. LA RUE 69

Vol. V. Sources of the Synoptic Gospels. By Rev. Dr. Carl S. Patton, First Congregational Church, Los Angeles, California. Pp. xiii + 263. $1.30 net.

Size, 28 × 18.5 cm. 4to.

Vol. VI. Athenian Lekythoi with Outline Drawing in Glaze Varnish on a White Ground. By Arthur Fairbanks, Director of the Museum of Fine Arts, Boston. With 15 plates, and 57 illustrations in the text. Pp. viii + 371. Bound in cloth. $4.00 net.

Vol. VII. Athenian Lekythoi with Outline Drawing in Matt Color on a White Ground, and an Appendix: Additional Lekythoi with Outline Drawing in Glaze Varnish on a White Ground. By Arthur Fairbanks. With 41 plates. Pp. x + 275. Bound in cloth $3.50 net.

Vol. VIII. The Old Testament Manuscripts in the Freer Collection. By Henry A. Sanders, University of Michigan. With 9 plates showing pages of the Manuscripts in facsimile. Pp. viii + 357. Bound in cloth. $3.50 net.

Parts Sold Separately in Paper Covers:

Part I. The Washington Manuscript of Deuteronomy and Joshua. With 3 folding plates. Pp. vi + 104. $1.25 net.

Part II. The Washington Manuscript of the Psalms. With 1 single plate and 5 folding plates. Pp. viii + 105–349. $2.00 net.

Vol. IX. The New Testament Manuscripts in the Freer Collection. By Henry A. Sanders, University of Michigan. With 8 plates showing pages of the Manuscripts in facsimile. Pp. x + 323. Bound in cloth. $3.50.

Parts Sold Separately in Paper Covers:

Part I. The Washington Manuscript of the Four Gospels. With 5 plates. Pp. vii + 247. $2.00 net.

Part II. The Washington Manuscript of the Epistles of Paul. With 3 plates. Pp. ix + 251–315. $1.25 net.

Vol. X. The Coptic Manuscripts in the Freer Collection. By William H. Worrell, Hartford Seminary Foundation. With 12 plates. Pp. xxvi + 396. Bound in cloth. $4.75 net.

Parts Sold Separately in Paper Covers:

Part I. The Coptic Psalter. The Coptic text in the Sahidic Dialect, with an Introduction, and with 6 plates showing pages of the Manuscript and Fragments in facsimile. Pp. xxvi + 112. $2.00 net.

Part II. A Homily on the Archangel Gabriel by Celestinus, Bishop of Rome, and a Homily on the Virgin by Theophilus, Archbishop of Alexandria, from Manuscript Fragments in the Freer Collection and the British Museum. The Coptic Text with an Introduction and Translation, and with 6 plates showing pages of the Manuscripts in facsimile. Pp. 113–396. $2.50 net.

THE MACMILLAN COMPANY

Publishers 64–66 Fifth Avenue New York

Part II. MUSIC TO THE ALCESTIS OF EURIPIDES WITH ENGLISH TEXT.

Part III. MUSIC TO THE IPHIGENIA AMONG THE TAURIANS BY EURIPIDES, WITH GREEK TEXT.

Part IV. TWO FRAGMENTS OF ANCIENT GREEK MUSIC.

Part V. MUSIC TO CANTICA OF THE MENAECHMI OF PLAUTUS.

Part VI. ATTIS: A SYMPHONIC POEM.

VOL. XVI. NICOMACHUS OF GERASA: INTRODUCTION TO ARITHMETIC. Translated into English by Martin Luther D'Ooge, with Studies in Greek Arithmetic by Frank Egleston Robbins and Louis C. Karpinski. (*In preparation.*)

VOLS. XVII, XVIII. CORRESPONDENCE OF THE ASSYRIAN KINGS: SARGON PERIOD. Translated into English, with a transliteration of the TEXT AND A COMMENTARY. By Leroy Waterman, University of Michigan. (*In press.*)

FACSIMILES OF MANUSCRIPTS

Size, 40.5 × 35 cm.

FACSIMILE OF THE WASHINGTON MANUSCRIPT OF DEUTERONOMY AND JOSHUA IN THE FREER COLLECTION. With an Introduction by Henry A. Sanders. Pp. x; 201 heliotype plates. The University of Michigan. Ann Arbor, Michigan, 1910.

Limited edition, distributed only to Libraries, under certain conditions. A list of Libraries containing this Facsimile is printed in *University of Michigan Studies, Humanistic Series*, Volume VIII, pp. 351–353.

Size, 34 × 26 cm.

FACSIMILE OF THE WASHINGTON MANUSCRIPT OF THE FOUR GOSPELS IN THE FREER COLLECTION. With an Introduction by Henry A. Sanders. Pp. x; 372 heliotype plates and 2 colored plates. The University of Michigan. Ann Arbor, Michigan, 1912.

Limited edition, distributed only to Libraries, under certain conditions. A list of Libraries containing this Facsimile is printed in *University of Michigan Studies, Humanistic Series*, Volume IX, pp. 317–320.

SCIENCE SERIES

Size, 28 × 18.5 cm. 4°. Bound in Cloth.

VOL. I. THE CIRCULATION AND SLEEP. By John F. Shepard, University of Michigan. Pp. ix + 83, with an Atlas of 63 plates, bound separately. Text and Atlas, $2.50 net.

VOL. II. STUDIES ON DIVERGENT SERIES AND SUMMABILITY. By Walter B. Ford, University of Michigan. Pp. xi + 194. $2.50.

THE MACMILLAN COMPANY

Publishers 64–66 Fifth Avenue New York

Printed and bound by CPI Group (UK) Ltd, Croydon, CR0 4YY

16/09/2026

14963492-0001